Genetics, Genomics and Breeding of Peanuts

Genetics, Genomics and Breeding of Peanuts

Editor

Mr. Shivaji Shende and Mr. Pankaj K. Dubey

scitus
academics

Genetics, Genomics and Breeding of Peanuts

Edited by **Shivaji Shende and Pankaj K. Dubey**

ISBN: 978-1-68117-025-1

Library of Congress Control Number: 2015931526

SCITUS Academics LLC,
www.scitusacademics.com
Box No. 4766, 616 Corporate Way,
Suite 2, Valley Cottage,
NY 10989

Printed in United States of America on Acid Free Paper ♾

Contents

Preface vii

Chapter 1 **Peanut Allergy, Allergen Composition, and Methods of Reducing Allergenicity: A Review** 1
Yang Zhou, Jin-shui Wang, Xiao-jia Yang, Dan-hua Lin, Yun-fang Gao, Yin-jie Su, Sen Yang, Yan-jie Zhang, and Jing-jing Zheng

Chapter 2 **Genetic Engineering Peanut for Higher Droughtand Salt-Tolerance** 23
Li Sun, Rongbin Hu, Guoxin Shen, Hong Zhang

Chapter 3 **Impact of Molecular Genetic Research on Peanut Cultivar Development** 41
C. Corley Holbrook, Peggy Ozias-Akins, Ye Chu and Baozhu Guo

Chapter 4 **Mutagenesis: A Useful Tool for the Genetic Improvement of the Cultivated Peanut (Arachis hypogaea L.)** 65
Chuan Tang Wang, Yue Yi Tang, Xiu Zhen Wang, Qi Wu, Hua Yuan Gao, Tong Feng, Jun Wei Su, Shu Tao Yu, Xian Lan Fang, Wan Li Ni, Yan Sheng Jiang, Lang Qian and Dong Qing Hu

Chapter 5 **Arachis Batizocoi: A Study of its Relationship to Cultivated Peanut (A. hypogaea) and its Potential for Introgression of wild genes into the Peanut crop using Induced Allotetraploids** 81
Soraya C. M. Leal-Bertioli, Silvio P. Santos, Karinne M. Dantas, Peter W. Inglis, Stephan Nielen, Ana C. G. Araujo, Joseane P. Silva, Uiara Cavalcante, Patricia M. Guimarães, Ana Cristina M. Brasileiro, Noelia Carrasquilla-Garcia, R. Varma Penmetsa, Douglas Cook, Márcio C. Moretzsohn and David J. Bertioli

Chapter 6 **Peanut (Arachis Hypogaea) Expressed Sequence Tag Project: Progress and Application** 117
Suping Feng Xingjun Wang, Xinyou Zhang, Phat M. Dang, C. Corley Holbrook, Albert K. Culbreath, Yaoting Wu, and Baozhu Guo

Chapter 7 **Groundnut Improvement: Use of Genetic and Genomic Tools......143**

Pasupuleti Janila, S. N. Nigam, Manish K. Pandey, P. Nagesh and Rajeev K. Varshney

Chapter 8 **Kazusa Marker DataBase: A Database for Genomics,Genetics, and Molecular breeding in Plants..191**

Kenta Shirasawa, Sachiko Isobe, Satoshi Tabata, and Hideki Hirakawa

Chapter 9 **Pre-Breeding for Diversification of Primary gene pool and Genetic Enhancement of Grain Legumes 213**

Shivali Sharma, H. D. Upadhyaya, R. K. Varshney, and C. L. L. Gowda

Chapter 10 **Construction of a SNP-based Genetic Linkage map in Cultivated Peanut based on large scale Marker Development using next-Generation Double-Digest Restriction-site-Associated DNA Sequencing (ddRADseq)..255**

Xiaojing Zhou, Youlin Xia, Xiaoping Ren, Yulin Chen, Li Huang, Shunmou Huang, Boshou Liao, Yong Lei, Liyin Yan, and Huifang Jiang

Citations..285

Index..289

Preface

Peanuts have substantial nutrients and have been consumed in various forms around the world for a long time. Peanut butter is only one such product which has been consumed in large quantities, especially in Western countries since many years. It is not so well-known that the domestic market is dominated by milk butter. Peanuts are filled with other nutrients and unsaturated fats that can improve glucose and homeostasis of insulin. Also it can help reduce diabetes. It can be used as a replacement for processed or red meats. It's very popular in western countries. Peanut butter is an excellent milk butter substitute. It's a high protein low calorie product. The price tag is unbelievably competitive compared to milk butter. Peanut butter is most commonly used as a spread in sandwiches. It is popular in cakes and bakeries. Other applications include cookies, frosting, cakes, salads, cupcakes, pies, celery stuffed, muffins, biscuits, apples, soup, peanut butter rolls, boiled custard, meat loaf, maize bread, vegetable sauces, brownies, sweet potatoes, etc. It is also used in chocolate and ice cream. It is also used in the USA, together with other Western countries, as table butter in calories lower than butter milk. Integrating comprehensive and up-to-date information on genomic and genetic resources, and molecular breeding techniques, this book will serve as a valuable tool for students and researchers on the latest information on the genetics and genomics of peanuts.

Chapter 1

Peanut Allergy, Allergen Composition, and Methods of Reducing Allergenicity: A Review

Yang Zhou,[1] Jin-shui Wang,[2] Xiao-jia Yang,[1] Dan-hua Lin,[1] Yun-fang Gao,[1] Yin-jie Su,[1] Sen Yang,[2] Yan-jie Zhang,[2] and Jing-jing Zheng[1]

[1]College of Food Science and Technology, Henan University of Technology, Zhengzhou 450001, China

[2]College of Bioengineering, Henan University of Technology, Zhengzhou 450001, China

ABSTRACT

Peanut allergy affects 1-2% of the world's population. It is dangerous, and usually lifelong, and it greatly decreases the life quality of peanut-allergic individuals and their families. In a word, peanut allergy has become a major health concern worldwide. Thirteen peanut allergens are identified, and they are briefly introduced in this paper. Although there is no feasible solution to peanut allergy at present, many methods have shown great promise. This paper reviews methods of reducing peanut allergenicity, including physical methods (heat and pressure, PUV), chemical methods (tannic acid and magnetic beads), and biological methods (conventional breeding, irradiation breeding, genetic engineering, enzymatic treatment, and fermentation).

INTRODUCTION

Food allergy is a worldwide health problem. It affects approximately 5% of young children and 3% to 4% of adults in westernized countries [1], and it becomes more and more common in developing countries. Although virtually any food can cause allergy, over 90% of food allergy is triggered by eight food sources: milk, egg, peanut, tree nuts, shellfish, fish, wheat, and soy [2]. Among them, peanut is one of the most allergenic. Peanut allergy affects many individuals and its prevalence is increasing rapidly (the prevalence of peanut allergy in some countries is summarized in Table 1). In western countries, the prevalence of peanut allergy in children in the USA increased from 0.4% in 1997 to 1.4% in 2008 [3]; the prevalence of sensitization to peanuts of 3-year olds in the UK rose from 1.3% to 3.2% between 1989 and 1995 [4]; over 1% of Canadian children are allergic to peanuts [5]; the prevalence of peanut allergy in Denmark and France is 0.2–0.4% and 0.3–0.75%, respectively [6, 7]. In Asia, although few epidemic studies of peanut allergy have been carried out, a study suggests that 0.47% of 14–16-year-old local Singapore schoolchildren and 0.43% of 14–16-year-old Philippine schoolchildren are allergic to peanuts [8]. Considering that 76.8% of Singapore residents are Chinese [9], peanut allergy is likely to become serious in China in the future. Moreover, peanut allergy can sometimes be life-threatening and usually cannot be outgrown, and it is almost impossible to avoid accidental ingestion [10,11]. Therefore, peanut allergy greatly reduces the life quality of the patient [12] and brings trouble to food industry in allergen labeling. Solving this problem has a great significance not only to the peanut-allergic individuals but also to the food industry.

Table 1. The prevalence of peanut allergy in some countries [3–8]

Countries	Prevalence
US children	1.40%
Britain children	3.2%
Canadian children	1.03%
Denmark	0.2–0.4%
France	0.3–0.75%
Local Singapore schoolchildren (14–16 years old)	0.47%
Philippine schoolchildren (14–16 years old)	0.43%

PEANUT ALLERGEN

To date, 13 peanut allergens (Ara h 1 through h 13) have been recognized by the Allergen Nomenclature Sub-Committee of the International Union of Immunological Societies. These allergens come from 7 protein families. Except for Ara h 1 (150 kD) and Ara h 3 (360–380 kD), the molecular weight of the other allergens ranges from 5 to 17 kD [13]. The genes corresponding to the 13 allergens have already been elucidated, the sequence of many linear epitopes of peanut allergens has been identified (Table 2), and the 3D models of Ara h 1-Ara h 6 have been built. Although the allergenicity of these allergens has not been thoroughly studied and there is still some debate about the definition of major allergens, the major peanut allergens that are most widely accepted are Ara h 1, Ara h 2, and Ara h 3.

Table 2. Sequence of linear epitopes of peanut allergens [17, 19, 26, 30]

Allergen	Epitope number	Epitope sequence
Ara h 1 core region	7[a]	PGQFEDFF
	8[a]	YLQGFSRN
	9[a]	FNAEFNEIRR
	10[a]	QEERGQRR
	11[a]	DITNPINLRE
	12[a]	NNFGKLFEVK
	13[a]	GNLELV
	14[a]	RRYTARLKEG
	15[a]	ELHLLGFGIN
	16[a]	HRIFLAGDKD
	17[a]	IDQIEKQAKD
	18[a]	KDLAFPGSGE
	19[a]	KESHFVSARP
	21[b]	NEGVIVKVSKEHVEELTKHAKSVSK
Ara h 2	1	HASARQQWEL
	2	QWELQGDRRC
	3	DRRCQSQLER
	4	LRPCEQHLMQ
	5	KIQR.DEDSYE
	6	YERDPYSPSQ
	7	SQDPYSPSPY
	8	DRLQ..GRQQEQ
	9	KRELRNLPQQ
	10	QRCDLDVESG
Ara h 3	1	IETWNPNNQEFECAG
	2	GNIFSGFTPEFLAQA
	3	VTVRGGLRILSPDRK
	4	DEDEYEYDE--EDRRRG

aDetermined by [17]. bDetermined by [19].

Ara h 1

Ara h 1 is a glycoprotein and belongs to the vicilin (7S) family. It comprises 12–16% of the total peanut protein [14] and affects 35–95% of peanut-allergic patients in different populations [15]. Native Ara h 1 exists as a trimer formed by three identical monomers, and the crystal structure of its core region has been elucidated (Figure 1) [16]. The topology and basic structure of its core region are very similar to other known structures of 7S globulins. Those similarities indicate that there is a high possibility of cross-reactivity between 7S globulins [16]. To date, 21 linear epitopes have been identified on the mature Ara h 1 [17–20], and 14 epitopes were found in the core region [16]. It is found that most epitopes on the core region become either slightly (<50% burial) or significantly (≥50% burial) buried upon trimer formation [16]. The burial of those epitopes likely explains the relatively weak activity of native (trimer) Ara h 1 in cross-linking IgE and the strong binding of IgE to denatured monomers [18, 21].

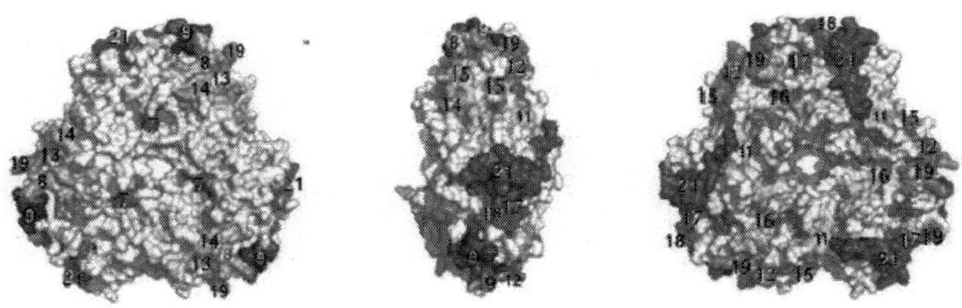

Figure 1. IgE epitopes are mapped on the surface of the 3D model of Ara h 1 core region [16].

Ara h 2

Ara h 2 (16-17 kDa) is also a glycoprotein and accounts for 5.9–9.3% of the total peanut protein [22]. It is a 2S albumin, also known as conglutin, and functions as a trypsin inhibitor [23]. More than 95% of peanut-allergic individuals in the USA have specific IgE to Ara h 2, and Ara h 2 was found to be a more potent allergen than Ara h 1 [21, 24, 25]. The structure of Ara h 2 is five α-helices arranged in a right-handed superhelix and connected by several extended loops (Figure 2). This three-dimensional conformation is stabilized by four conserved disulphide bridges. Ten epitopes have been mapped on Ara h 2, and these epitopes show a fairly well exposition on the molecular surface [26].

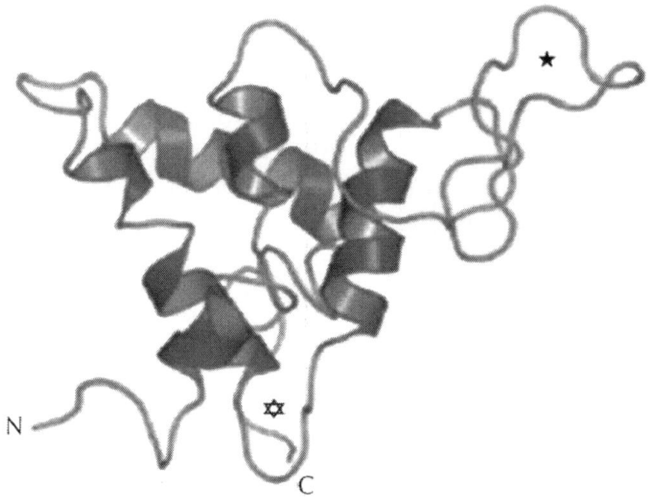

Figure 2. Ribbon diagram of Ara h 2 [26].

Ara h 3

Ara h 3 is a seed storage protein and belongs to the legumin (11S) family [27]. It is recognized by 50% of peanut-allergic individuals and also functions as a trypsin inhibitor [28, 29]. Ara h 3 and soybean glycinin result in a sequence identity of 47.2% [30]. Mature Ara h 3 is a hexamer (360–380 kD) formed by a head-to-head association of two trimers (Figure 3) [30]. Each monomer was found to have 4 linear epitopes [31]. In the natural form of Ara h 3, epitope 4 is fully exposed, while the side chains of most of the critical residues of the other three epitopes are completely or nearly completely buried. This suggests that linear epitopes 1 and 2 may not be recognized by IgE antibodies in the intact form, while epitope 4 and part of epitope 3 may be allergic in the natural form of Ara h 3 [30].

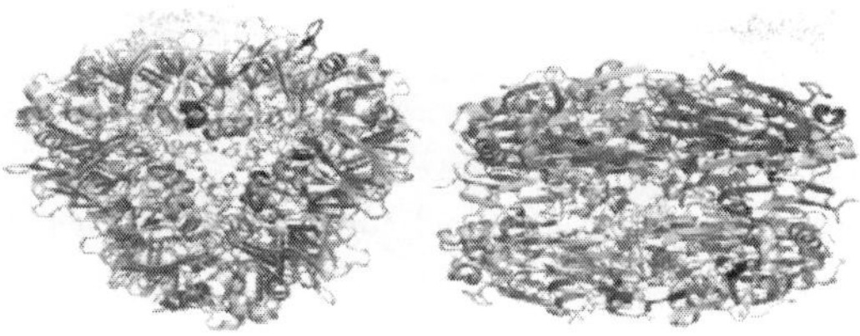

Figure 3. Ara h 3 is represented with each of the monomers shown in a different color. In the gray monomer, linear epitope 1, 2, and 3 are shown in red, green, and blue, respectively [30].

Ara h 4

Ara h 4 is actually an isoform of Ara h 3. Now, it is no longer thought to be a distinct allergen and renamed to Ara h 3.02 [13, 32].

Ara h 5

Ara h 5 (15 kD) belongs to the profilin family and regulates the polymerization of actin [13, 32]. It is presented at low levels in peanut extracts and is recognized by 13% of 40 patients' sera [27, 33]. The structure of Ara h 5 is shown in Figure 4.

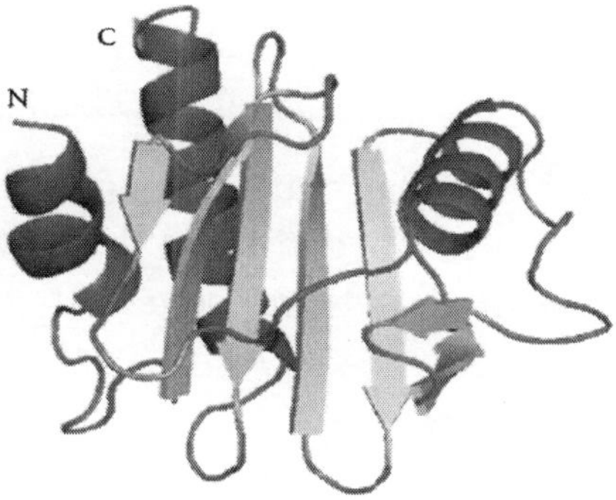

Figure 4. Ribbon diagram of the three-dimensional model of Ara h 5. Strands of b-sheet and stretches of a-helix are in yellow and red, respectively. Coil structures or loops are in green, N and C indicate the N- and C-terminus of the polypeptide, respectively [38].

Ara h 6

Ara h 6 is a 15 kD protein and belongs to the conglutin family [13]. It is 59% homologous to Ara h 2 and has similar allergenicity [34, 35]. Ara h 6 is a heat and digestion stable protein and showed resistance to proteolytic treatment [36, 37]. The structure of Ara h 6 is shown in Figure 5.

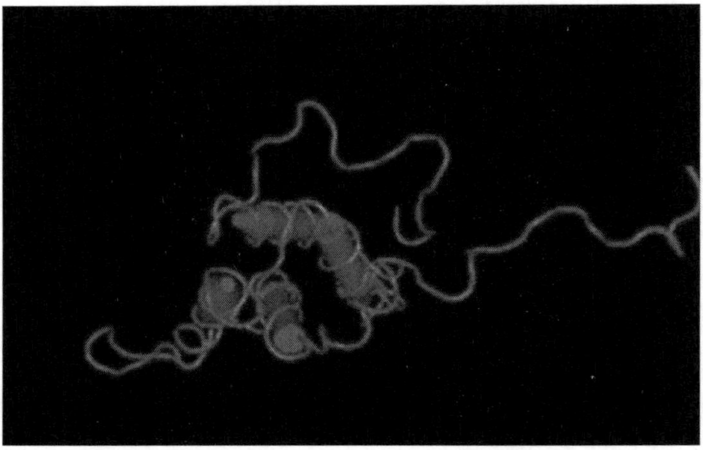

Figure 5. Ara h 6 (PDB Entry 1W2Q, first molecule in the entry).

Ara h 7

Ara h 7 is also a 15 kD protein and belongs to the conglutin family [13]. The sequence identity between Ara h 2 and Ara h 6 is 35%, and it is recognized by 13% of 40 patients' sera [23].

Other Peanut Allergens

Ara h 8 (17 kD) is a Pathogenesis-related protein. Ara h 9 (9.8 kD, 2 isoforms) is a nonspecific lipid-transfer protein. Ara h 10 (16 kD, 2 isoforms) and Ara h 11 (14 kD) belong to oleosin. Ara h 12 and Ara h 13 are defensin, with molecular weight ranging from 5 to 12 kD [13].

HARM OF PEANUT ALLERGY

Peanut-allergic reactions involve the skin, the respiratory tract, and the gastrointestinal tract [39]. The common symptoms include acute urticaria, acute vomiting, laryngeal oedema, hypotension, and dysrhythmia [40, 41]. Peanut allergy is very dangerous. Ingestion of even a trace amount of peanut may elicit life-threatening reactions within minutes [42]. Peanut, together with tree nuts, causes most of the fatal or near-fatal food-related anaphylaxis, and peanut allergy leads to 100–200 deaths each year in the USA [43, 44]. Moreover, peanut allergy is usually lifelong, with only 10% of peanut-allergic children outgrowing it [10]. Last but not least, due to the ubiquitous use of peanut in food industry, it is almost impossible for a peanut-allergic patient to completely avoid peanut even if he/she strictly obeys the doctor's guidance. Studies suggest that up to 75% of individuals with known peanut allergy experience reactions caused by accidental exposure [11]. Thus, peanut allergy gives enormous pressure to the patients and their families and greatly impairs their life quality [12]. In addition, the US law demands that allergen content be labeled on any product sold in the USA [45], and tracing and determining peanut allergens in food products increase the cost and bring inconvenience to the food trade.

METHODS OF REDUCING ALLERGENICITY

Although there is now no feasible solution to peanut allergy, many methods have shown great prospect, including oral immunotherapy and some methods to reduce peanut allergenicity. The methods of reducing peanut allergic potential are reviewed as follows.

Physical Methods

Heat and Pressure Treatment

There are three ways to decrease peanut allergenicity by heat treatment. The first is roasting. Roasting has been recognized as a process that can increase peanut allergenicity [46]. However, Vissers et al. found that after heating Ara h 2/6 (purified from raw peanuts) in a dry form for 20 min at 145°C, the IgE-binding capacity and the degranulation capacity of Ara h 2/6 were 600–700-fold lower than those in the native form [47].

The second is boiling. Boiling native Ara h 2/6 (15 min, 110°C) and boiling native Ara h 1 (15 min, 100°C) resulted in decreased IgE reactivity and mediator-releasing capacity; but for Ara h 2/6 and Ara h 1 extracted from roasted peanut, boiling had no effect [48, 49].

The third one is autoclaving. Cabanillas et al. discovered that IgE-binding capacity of peanut allergens is significantly decreased by autoclaving at 2.56 atm, for 30 min [50]. However, this method obviously comes with high energy consumption and expensive devices.

PUV

Pulsed ultraviolet light (PUV) is another effective method in reducing peanut allergenicity. Yang et al. treated protein extracts from raw and roasted peanuts and peanut butter slurry in a Xenon Steripulse XL 3000 PUV system. The treatment time was 2, 4, and 6 min for protein extracts and 1, 2, and 3 min for peanut butter slurry. The distance from the central axis of the lamp was varied at 10.8, 14.6, and 18.2 cm. The research found that PUV treatment resulted in reduction in the level of Ara h 1, Ara h 2, and Ara h 3 and decreased IgE binding ability by 12.9% to 6.7% [51]. However, like all the other irradiation technologies, this method comes with concern of food safety.

Chemical Methods

Tannic Acids

Chung and Reed reduced the allergenicity of peanut butter by adding tannic acid. The principal is that tannic acid interacts with allergens to form indigestible complex, and epitopes on the allergens are covered during complex formation, making the epitopes inaccessible to antibodies and resulting in reduced allergenicity. Chung and Reed added tannic acid to a peanut butter extract (5 mg/mL; pH = 7.2) and discovered that when pH = 2 and pH = 8, the complexes do not release Ara h 1, or Ara h 2, and the IgE binding ability is decreased; and when concentration of tannic acid is 1-2 mg/mL, the IgE binding ability of the complex is reduced substantially [52]. Since tannic acid interacts with both allergen and non-allergen peanut proteins, such treatment has two obvious deficiencies: first, peanut nutrition is reduced to a great extent, and second, intake of much indigestible food may cause stomach discomfort and thus greatly limit consumption of peanut products.

Magnetic Beads

Magnetic beads can also be used to remove peanut allergens. The principle is that phenolic compounds and ferric ions (Fe^{3+}) can bind to peanut allergens; thus, one can reduce peanut allergenicity by using magnetic beads attached with or without phenolics to capture peanut allergens or allergen-Fe^{3+} complexes and then separate the beads by a magnetic device. Chung and Champagne found the following: treating peanut extracts by CHL beads (magnetic beads covalently attached with chlorogenic acid, a phenolic) resulted in marked decrease of Ara h 1 and small reduction of Ara h 2; when using magnetic beads without phenolic compounds to treat peanut extracts that have been incubated with Fe^{3+} and dialyzed, both Ara h 1 and Ara h 2 were markedly reduced; those two methods reduced IgE binding ability of the treated extracts by 28–47%. Chung and Champagne believed that the magnetic beads system was a simple way to partially remove peanut allergens from peanut extracts, and it could be a potential approach to produce hypoallergenic peanut products and beverages [53].

Biological Methods

Conventional Breeding

The rationale of conventional breeding is crossing hypoallergenic varieties to produce a variety that is more hypoallergenic. Perkins et al. crossbred peanuts that were missing either an Ara h 2 or Ara h 3 isoform and produced a variety lacking both isoforms. The observed numbers of the new variety conformed to the 15 : 1 Mendelian dihybrid ratio [54]. However, considering the large amount of peanut allergens, the progress of this method seems very slow.

Irradiation Breeding

As for mutation breeding, a type of technology is well worth mentioning. It is the heavy-ion beam irradiation (HIBI). This technology leads to mutation and inactivation of a single gene or multiple genes in a plant, thus inducing stable knockout mutants [55, 56]. Cabanos et al. treated a Japanese peanut variety—Nakateyutaka—with either N or C heavy-ion beams at a dose of 100 Gy and obtained seventeen knockout mutants from 11,335 screened M2 seeds. Among the seventeen mutants, eight lacked either one of the two isoforms of Ara h 2, and the other nine are missing one of the isoforms of Ara h 3 [57]. Cabanos et al. believe that HIBI is a powerful means of producing knockout hypoallergenic peanuts and has many advantages [57], including low radiation exposure levels, less cellular damage, no need for tedious tissue culture or regenerative procedures, no severe growth inhibition, and, in general, less plant death and a high rate of mutation producing diverse kinds of mutants [55, 56, 58]. However, like all the other irradiation technologies, HIBI comes with the concern of food safety.

Genetic Engineering

Great advance has been made in removing peanut allergens by genetic technology. Chu et al. silenced Ara h 2 and Ara h 6 by RNA interference and produced three independent transgenic lines. All the three lines were featured by significant reduction in human IgE binding to Ara h 2 and Ara h 6 as well as the level of Ara h 2, whereas the level of Ara h 6 was only reduced in two lines. In addition, there were no significant differences between the seed weight and germination data of transgenic and nontransgenic plants [59]. Another research comes from Ananga et al. who tried to produce hypoallergenic peanuts by silencing Ara h 1, Ara h 2, and Ara h 3 with RNA interference. Ananga et al. have found the following: the percentages of transgenic peanut that showed reduction in Ara h 1, Ara h 2, and Ara h 3 were 9%, 10%, and 16%, respectively 3% transgenic seeds were free of all three allergens; the IgE-binding capacity was significantly reduced in at least nine transgenic seeds with reduction in Ara h 1 or Ara h 2, Ara h 3 [60].

Although genetic technology has shown great promise to produce allergen-free peanut, this technology also has two big drawbacks. One is people's increasing repulsion to transgenic food. The other is the fact that peanut allergens account for 20–30% of total peanut proteins, and if all the allergens are removed, peanuts may not taste like peanuts.

Enzymatic Treatment

Enzymatic treatment is full of potential to produce allergen-free peanut and there are two types of enzymatic treatment.

One is using enzymes to cross-link allergen proteins, resulting in the burial of epitopes. Chung and Champagne treated protein extract from roasted and raw peanuts with peroxidase (POD) and transglutaminase (TGA) at 37°C. Both enzymes catalyze cross-links between proteins. Chung and Champagne found the following: POD treatment of roasted peanut resulted in partial loss of Ara h 1 and Ara h 2 along with reduced IgE binding ability and formation of new polymers; on the other hand, TGA treatment of roasted peanut had no effect on the content of Ara h 1 and Ara h 2 as well as IgE binding ability; both POD and TGA had no effect on the IgE binding ability of protein extract from raw peanut. Chung and Champagne believed that POD may be useful in desensitizing peanut while TGA should be useless [61].

The other is using enzymes to break down allergens, destroying their epitopes. Cabanillas et al. studied the effect of hydrolysis with alcalase and flavourzyme on the allergenicity of the soluble protein fraction of roasted peanut. Parameters for alcalase hydrolysis were S=2%, E/S=0.4 AU/g of protein, T=50°C, and pH 8.0; parameters for flavourzyme hydrolysis were S=2%, E/S=100 LAPU/g of protein, °C, and pH 7.0. Cabanillas et al. discovered the following: 30 min alcalase treatment resulted in an important decrease of Ara h 1, Ara h 2, and Ara h 3 levels and reduced IgE binding reactivity by 98%; 90 min alcalase treatment could fully eliminate IgE binding reactivity; while 30 min flavourzyme treatment caused an increase in IgE reactivity, hydrolyzing with flavourzyme for 300 min led to a 65% inhibition of IgE reactivity [62]. Although Cabanillas et al. demonstrated that enzymatic treatment with alcalase or flavourzyme could reduce IgE reactivity in peanuts, Guo et al. found that the allergenicity was retained after treating roasted peanut protein extract in a similar way [63]. The two researches adopted different methods to assess the allergenicity of enzymatic products, and the liability of such assessment methods is still in debate. It is unclear whether alcalase and flavourzyme have an effect or not.

The most promising enzymatic method for desensitizing peanut is the research of Ahmedna et al. Mohamed's team has been working on the subject for over 7 years. They have developed a method which is very likely to completely eliminate peanut allergenicity in a quick, simple, and inexpensive way, without greatly changing the flavor and texture of natural peanuts. Since Mohamed's team are using their research to apply a patent, only a little detail of the method can be obtained. It is only told that peanut allergenicity is reduced by direct application of enzymatic solution to either raw, blanched, or roasted peanuts, or peanut products or derivatives (including but not limited to peanut butter, peanut kernels, peanut skins, peanut protein isolate, peanut flour, or peanut milk); the enzymatic solution used in this method contains at least one endopeptidase whose hypoallergenically effective amount is at least 0.001% (w/w) [64].

Fermentation

Few studies have been reported on reducing peanut allergenicity by fermentation. It is only reported that Dr. Ahmedna et al. found that fermenting whole or ground peanuts with an edible fungus reduced the detectable level of major allergenic proteins Ara h 1 and Ara h 2 by as much as 70 percent, and this study is still in the early stages [65]. Although fermentation method has rarely been reported, this method has already successfully reduced the allergic potential of soybean meal and bovine whey proteins [66, 67] and is very likely to reduce the allergenicity of peanuts. The major principle of fermentation is almost the same as that of enzymatic treatment, and fermentation has all the merits of the enzyme method. Furthermore, this method is usually much cheaper. Therefore, fermentation is still a very promising method to produce hypoallergenic peanuts.

PROSPECT

Breeding is an effective way, but some problems cannot be ignored. The advance of conventional breeding seems very slow, and mutation breeding is always involved with the problem of food safety; all of these impede the development of the breeding method.

Heat and pressure treatment is another effective approach to reduce peanut allergenicity. However, it still has some deficiencies: roasting can only be an assistant method; the effect of boiling is very limited; autoclaving requires high energy consumption and expensive devices. All of these constitute obstacles to the development of this method.

Transgenic technology is very promising to produce allergen-free variety in the near future. However, with people's repulsion to transgenic food and flavor problems, it has a long way to go to have transgenic allergen-free peanuts in the market.

Tannic acid is a useful agent, but considering its obvious drawbacks, it can only be an assistant approach. PUV has the same problem of mutation breeding which hinders its development. Magnetic beads capture is a promising way to decrease peanut allergenicity.

At present, enzymatic treatment is the most promising way to produce nonallergic peanuts. Compared with breeding and gene technology, enzymatic treatment is mild, is natural, usually does not produce harmful substance and can be readily accepted by the public. Compared with autoclaving, the approach of Mohamed's team is very cheap. Moreover, enzymatic treatment does not impair peanut nutrition value. Therefore, the authors believe that it is the most promising approach nowadays.

Fermentation is full of potential to reduce peanut allergenicity. Moreover, it has all the merits of the enzyme method and is usually much cheaper. Although this method has rarely been reported, it may be the best way to reduce peanut allergenicity.

REFERENCES

1. S. H. Sicherer and H. A. Sampson, "Food allergy," Journal of Allergy and Clinical Immunology, vol. 125, supplement 2, pp. S116–S125, 2010.
2. S. L. Hefle, J. A. Nordlee, and S. L. Taylor, "Allergenic foods," Critical Reviews in Food Science and Nutrition, vol. 36, supplement 1, pp. S69–S89, 1996.

3. S. H. Sicherer, A. Muñoz-Furlong, J. H. Godbold, and H. A. Sampson, "US prevalence of self-reported peanut, tree nut, and sesame allergy: 11-year follow-up," Journal of Allergy and Clinical Immunology, vol. 125, no. 6, pp. 1322–1326, 2010.
4. J. Grundy, S. Matthews, B. Bateman, T. Dean, and S. H. Arshad, "Rising prevalence of allergy to peanut in children: data from 2 sequential cohorts," Journal of Allergy and Clinical Immunology, vol. 110, no. 5, pp. 784–789, 2002.
5. M. Ben-Shoshan, R. S. Kagan, R. Alizadehfar et al., "Is the prevalence of peanut allergy increasing? A 5-year follow-up study in children in Montreal," Journal of Allergy and Clinical Immunology, vol. 123, no. 4, pp. 783–788, 2009.
6. M. Morisset, D.-A. Moneret-Vautrin, and G. Kanny, "Prevalence of peanut sensitizion in a population of 4,737 subjects-an allergo-vigilance network enquiry carried out in 2002," European Annals of Allergy and Clinical Immunology, vol. 37, no. 2, pp. 54–57, 2005.
7. M. Osterballe, T. K. Hansen, C. G. Mortz, A. Høst, and C. Bindslev-Jensen, "The prevalence of food hypersensitivity in an unselected population of children and adults," Pediatric Allergy and Immunology, vol. 16, no. 7, pp. 567–573, 2005.
8. L. P. Shek, E. A. Cabrera-Morales, S. E. Soh et al., "A population-based questionnaire survey on the prevalence of peanut, tree nut, and shellfish allergy in 2 Asian populations," Journal of Allergy and Clinical Immunology, vol. 126, no. 2, pp. 324–331, 2010.
9. Central Intelligence Agency, Library, Publications, The World Factbook, Singapore, people and society, ethnic groups, 2000, https://www.cia.gov/library/publications/the-world-factbook/geos/sn.html.
10. Berger and Smith, "Science commentary: why do some children grow out of peanut allergy?" British Medical Journal, vol. 316, p. 1275, 1998.
11. R. S. Kagan, L. Joseph, C. Dufresne et al., "Prevalence of peanut allergy in primary-school children in Montreal, Canada," Journal of Allergy and Clinical Immunology, vol. 112, no. 6, pp. 1223–1228, 2003.

12. R. M. King, R. C. Knibb, and J. O. Hourihane, "Impact of peanut allergy on quality of life, stress and anxiety in the family," Allergy, vol. 64, no. 3, pp. 461–468, 2009.
13. "Allergen Nomenclature (IUIS Allergen Nomenclature Sub-Committee),"http://www.allergen.org/search.php?Allergensource=Arachis+hypogaea.
14. E. C. de Jong, M. Van Zijverden, S. Spanhaak, S. J. Koppelman, H. Pellegrom, and A. H. Penninks, "Identification and partial characterization of multiple major allergens in peanut proteins," Clinical and Experimental Allergy, vol. 28, no. 6, pp. 743–751, 1998.
15. A. Mari, E. Scala, P. Palazzo, S. Ridolfi, D. Zennaro, and G. Carabella, "Bioinformatics applied to allergy: allergen databases, from collecting sequence information to data integration. The Allergome platform as a model," Cellular Immunology, vol. 244, no. 2, pp. 97–100, 2006.
16. C. Cabanos, H. Urabe, M. R. Tandang-Silvas, S. Utsumi, B. Mikami, and N. Maruyama, "Crystal structure of the major peanut allergen Ara h 1," Molecular Immunology, vol. 49, no. 1-2, pp. 115–123, 2011.
17. A. W. Burks, D. Shin, G. Cockrell, J. S. Stanley, R. M. Helm, and G. A. Bannon, "Mapping and mutational analysis of the IgE-binding epitopes on Ara h 1, a legume vicilin protein and a major allergen in peanut hypersensitivity," European Journal of Biochemistry, vol. 245, no. 2, pp. 334–339, 1997.
18. D. S. Shin, C. M. Compadre, S. J. Maleki et al., "Biochemical and structural analysis of the IgE binding sites on Ara h1, an abundant and highly allergenic peanut protein," Journal of Biological Chemistry, vol. 273, no. 22, pp. 13753–13759, 1998.
19. W. G. Shreffler, K. Beyer, T. T. Chu, A. W. Burks, and H. A. Sampson, "Microarray immunoassay: association of clinical history, in vitro IgE function, and heterogeneity of allergenic peanut epitopes,"Journal of Allergy and Clinical Immunology, vol. 113, no. 4, pp. 776–782, 2004.
20. H. J. Wichers, T. De Beijer, H. F. J. Savelkoul, and A. Van Amerongen, "The major peanut allergen Ara h 1 and its cleaved-off N-terminal peptide; possible implications for peanut allergen

detection," Journal of Agricultural and Food Chemistry, vol. 52, no. 15, pp. 4903–4907, 2004.

21. G. W. Palmer, D. A. Dibbern Jr., A. W. Burks et al., "Comparative potency of Ara h 1 and Ara h 2 in immunochemical and functional assays of allergenicity," Clinical Immunology, vol. 115, no. 3, pp. 302–312, 2005.
22. S. J. Koppelman, R. A. A. Vlooswijk, L. M. J. Knippels et al., "Quantification of major peanut allergens Ara h 1 and Ara h 2 in the peanut varieties Runner, Spanish, Virginia, and Valencia, bred in different parts of the world," Allergy, vol. 56, no. 2, pp. 132–137, 2001.
23. S. J. Maleki, O. Viquez, T. Jacks et al., "The major peanut allergen, Ara h 2, functions as a trypsin inhibitor, and roasting enhances this function," Journal of Allergy and Clinical Immunology, vol. 112, no. 1, pp. 190–195, 2003.
24. A. M. Scurlock and A. W. Burks, "Peanut allergenicity," Annals of Allergy, Asthma and Immunology, vol. 93, supplement 5, pp. S12–S18, 2004.
25. S. J. Koppelman, M. Wensing, M. Ertmann, A. C. Knulst, and E. F. Knol, "Relevance of Ara h1, Ara h2 and Ara h3 in peanut-allergic patients, as determined by immunoglobulin E Western blotting, basophil-histamine release and intracutaneous testing: Ara h2 is the most important peanut allergen," Clinical and Experimental Allergy, vol. 34, no. 4, pp. 583–590, 2004.
26. A. Barre, J. Borges, R. Culerrier, and P. Rougé, "Homology modelling of the major peanut allergen Ara h 2 and surface mapping of IgE-binding epitopes," Immunology Letters, vol. 100, no. 2, pp. 153–158, 2005.
27. S. J. Koppelman, E. F. Knol, R. A. A. Vlooswijk et al., "Peanut allergen Ara h 3: isolation from peanuts and biochemical characterization," Allergy, vol. 58, no. 11, pp. 1144–1151, 2003.
28. H. W. Wen, W. Borejsza-Wysocki, T. R. DeCory, and R. A. Durst, "Peanut allergy, peanut allergens, and methods for the detection of peanut contamination in food products," Comprehensive Reviews in Food Science and Food Safety, vol. 6, no. 2, pp. 47–58, 2007.
29. H. W. Dodo, O. M. Viquez, S. J. Maleki, and K. N. Konan, "cDNA clone of a putative peanut (Arachis hypogaea L.) trypsin

inhibitor has homology with peanut allergens Ara h 3 and Ara h 4," Journal of Agricultural and Food Chemistry, vol. 52, no. 5, pp. 1404–1409, 2004.

30. T. C. Jin, A. F. Guo, Y. W. Chen, A. Howard, and Y. Zhang, "Crystal structure of Ara h 3, a major allergen in peanut," Molecular Immunology, vol. 46, no. 8-9, pp. 1796–1804, 2009.
31. P. Rabjohn, E. M. Helm, J. S. Stanley et al., "Molecular cloning and epitope analysis of the peanut allergen Ara h 3," Journal of Clinical Investigation, vol. 103, no. 4, pp. 535–542, 1999.
32. H. Breiteneder and C. Radauer, "A classification of plant food allergens," Journal of Allergy and Clinical Immunology, vol. 113, no. 5, pp. 821–830, 2004.
33. T. Kleber-Janke, R. Crameri, U. Appenzeller, M. Schlaak, and W. Becker, "Selective cloning of peanut allergens, including profilin and 2S albumins, by phage display technology," International Archives of Allergy and Immunology, vol. 119, no. 4, pp. 265–274, 1999.
34. S. J. Koppelman, G. A. H. De Jong, M. Laaper-Ertmann et al., "Purification and immunoglobulin E-binding properties of peanut allergen Ara h 6: evidence for cross-reactivity with Ara h 2," Clinical and Experimental Allergy, vol. 35, no. 4, pp. 490–497, 2005.
35. X. Chen, Q. Wang, R. El-Mezayen, et al., "Ara h 2 and Ara h 6 have similar allergic effector activity and are substantially redundant," International Archives of Allergy and Immunology, vol. 160, no. 3, pp. 251–258, 2013.
36. M. Suhr, D. Wicklein, U. Lepp, and W. Becker, "Isolation and characterization of natural Ara h 6: evidence for a further peanut allergen with putative clinical relevance based on resistance to pepsin digestion and heat," Molecular Nutrition and Food Research, vol. 48, no. 5, pp. 390–399, 2004.
37. K. Lehmann, K. Schweimer, G. Reese et al., "Structure and stability of 2S albumin-type peanut allergens: implications for the severity of peanut allergic reactions," Biochemical Journal, vol. 395, no. 3, pp. 463–472, 2006.
38. C. Cabanos, M. R. Tandang-Silvas, V. Odijk et al., "Expression, purification, cross-reactivity and homology modeling of peanut

profilin," Protein Expression and Purification, vol. 73, no. 1, pp. 36–45, 2010.

39. S. H. Sicherer, A. W. Burks, and H. A. Sampson, "Clinical features of acute allergic reactions to peanut and tree nuts in children," Pediatrics, vol. 102, no. 1, p. e6, 1998.

40. S. A. Bock, A. Muoz-Furlong, and H. A. Sampson, "Fatalities due to anaphylactic reactions to foods,"Journal of Allergy and Clinical Immunology, vol. 107, no. 1, pp. 191–193, 2001.

41. H. A. Sampson, L. Mendelson, and J. P. Rosen, "Fatal and near-fatal anaphylactic reactions to food in children and adolescents," New England Journal of Medicine, vol. 327, no. 6, pp. 380–384, 1992.

42. H. A. Sampson, "Peanut allergy," New England Journal of Medicine, vol. 346, no. 17, pp. 1294–1299, 2002.

43. P. L. Jackson, "Peanut allergy: an increasing health risk for children," Pediatric Nursing, vol. 28, no. 5, pp. 496–504, 2002.

44. C. W. Lee and A. L. Sheffer, "Peanut allergy," Allergy and Asthma Proceedings, vol. 24, no. 4, pp. 259–264, 2003.

45. USDA, "Get the Facts: New Food Allergen Labeling Laws," 2006,http://www.fns.usda.gov/fdd/facts/nutrition/FoodAllergenFactSheet.pdf.

46. R. A. Kopper, N. J. Odum, M. Sen, R. M. Helm, J. S. Stanley, and A. W. Burks, "Peanut protein allergens: the effect of roasting on solubility and allergenicity," International Archives of Allergy and Immunology, vol. 136, no. 1, pp. 16–22, 2005.

47. Y. M. Vissers, M. Iwan, K. Adel-Patient et al., "Effect of roasting on the allergenicity of major peanut allergens Ara h 1 and Ara h 2/6: the necessity of degranulation assays," Clinical and Experimental Allergy, vol. 41, no. 11, pp. 1631–1642, 2011.

48. Y. M. Vissers, F. Blanc, P. S. Skov et al., "Effect of heating and glycation on the allergenicity of 2S albumins (Ara h 2/6) from peanut," PLoS ONE, vol. 6, no. 8, Article ID e23998, 2011.

49. F. Blanc, Y. M. Vissers, K. Adel-Patient et al., "Boiling peanut Ara h 1 results in the formation of aggregates with reduced allergenicity," Molecular Nutrition and Food Research, vol. 55, no. 12, pp. 1887–1894, 2011.

50. B. Cabanillas, S. J. Maleki, J. Rodríguez et al., "Heat and pressure treatments effects on peanut allergenicity," Food Chemistry, vol. 132, no. 1, pp. 360–366, 2012.
51. W. W. Yang, N. R. Mwakatage, R. Goodrich-Schneider, K. Krishnamurthy, and T. M. Rababah, "Mitigation of major peanut allergens by pulsed ultraviolet light," Food and Bioprocess Technology, vol. 5, no. 7, pp. 2728–2738, 2012.
52. S. Y. Chung and S. Reed, "Removing peanut allergens by tannic acid," Food Chemistry, vol. 134, no. 3, pp. 1468–1473, 2012.
53. S. Y. Chung and E. Champagne, "Using magnetic beads to reduce peanut allergens from peanut extracts," Journal of Allergy Clinical Immunology, vol. 125, no. 2, supplement 1, p. AB223, 2010.
54. T. Perkins, D. A. Schmitt, T. G. Isleib, et al., "Breeding a hypoallergenic peanut," The Journal of Allergy and Clinical Immunology, vol. 117, supplement 2, p. S328, 2006.
55. S. Kikuchi, Y. Saito, H. Ryuto et al., "Effects of heavy-ion beams on chromosomes of common wheat, Triticum aestivum," Mutation Research, vol. 669, no. 1-2, pp. 63–66, 2009.
56. Y. Kazama, H. Saito, M. Fujiwara et al., "An effective method for detection and analysis of DNA damage induced by heavy-ion beams," Bioscience, Biotechnology and Biochemistry, vol. 71, no. 11, pp. 2864–2869, 2007.
57. C. S. Cabanos, H. Katayama, H. Urabe et al., "Heavy-ion beam irradiation is an effective technique for reducing major allergens in peanut seeds," Molecular Breeding, vol. 30, no. 2, pp. 1037–1044, 2011.
58. T. Abe, T. Matsuyama, S. Sekido, I. Yamaguchi, S. Yoshida, and T. Kameya, "Chlorophyll-deficient mutants of rice demonstrated the deletion of a DNA fragment by heavy-ion irradiation," Journal of Radiation Research, vol. 43, pp. S157–S161, 2002.
59. Y. Chu, P. Faustinelli, M. L. Ramos et al., "Reduction of IgE binding and nonpromotion of Aspergillus flavus fungal growth by simultaneously silencing Ara h 2 and Ara h 6 in peanut," Journal of Agricultural and Food Chemistry, vol. 56, no. 23, pp. 11225–11233, 2008.
60. A. Ananga, H. Dodo, and K. Konan, "Elimination of the three major allergens in transgenic peanut (Arachis hypogea L),"

in Vitro Cellular & Developmental Biology-Animal, vol. 44, pp. S36–S37, 2008.

61. S. Chung and E. T. Champagne, "Effect of enzyme treatment on the allergenic properties of peanuts,"Journal of Allergy and Clinical Immunology, vol. 111, no. 2, 2003.

62. B. Cabanillas, M. M. Pedrosa, J. Rodríguez et al., "Influence of enzymatic hydrolysis on the allergenicity of roasted peanut protein extract," International Archives of Allergy and Immunology, vol. 157, no. 1, pp. 41–50, 2011.

63. R. Guo, X. Shi, B. White, et al., "Allergenicity of peanut proteins is retained following enzymatic hydrolysis," Journal of Allergy and Clinical Immunology, vol. 129, no. 2, p. AB367, 2011.

64. M. Ahmedna, J. M. Yu, and I. Goktepe, "Process for preparing hypoallergenic and non-allergenic peanut butter and associated products," 2010, United States Patent Application Publication, Pub. No: US2010/0080870 A1.

65. "New process removes allergy proteins from peanuts," Magazine of the Agricultural Research Program at North Carolina Agricultural and Technical State University, vol. 2, p. 3, 2005.

66. Y. S. Song, J. Frias, C. Martinez-Villaluenga, C. Vidal-Valdeverde, and E. G. de Mejia, "Immunoreactivity reduction of soybean meal by fermentation, effect on amino acid composition and antigenicity of commercial soy products," Food Chemistry, vol. 108, no. 2, pp. 571–581, 2008.

67. G. H. Bu, Y. K. Luo, Y. Zhang, and F. Chen, "Effects of fermentation by lactic acid bacteria on the antigenicity of bovine whey proteins," Journal of the Science of Food and Agriculture, vol. 90, no. 12, pp. 2015–2020, 2010.

Chapter 2

Genetic Engineering Peanut for Higher Droughtand Salt-Tolerance

Li Sun[1], Rongbin Hu[1], Guoxin Shen[2], Hong Zhang[1]

[1]Department of Biological Sciences, Texas Tech University, Lubbock, USA;

[2]Zhejiang Academy of Agricultural Sciences, Hangzhou, China

ABSTRACT

Peanut (Arachis hypogaea L.) is one of the major oilseed crops, mainly grown in tropical and sub-tropical regions of the world. It is also rich in proteins, vitamins and ions, therefore it constitutes an important portion of food nutrition for people in these regions. The production of peanut is being threatened by the changing environments as the major peanut producing counties such as China, India, and USA are facing severe water shortage for peanut irrigation. The yield and quality of peanut are negatively affected by drought and salinity. Making peanut more droughtand salt-tolerant will likely sustain peanut production in countries where water shortage or saline soil are already problems. Efforts were made to genetically engineer peanut for higher tolerance to drought and salt. Analysis of these transgenic peanut plants indicated that the agronomic traits such as peanut yields were the same between wild-type and transgenic peanut plants under normal growth conditions, yet the yields of transgenic peanut plants were much higher than wild-type peanut plant under reduced irrigation conditions. Other traits such as protein content and fatty acid compositions in the seeds of transgenic peanut plants were not altered under both normal and drought conditions, indicating that the genetic manipulation of peanut for stress tolerance did not affect chemical compositions of peanut seeds in transgenic peanut plants, only increased seed yields under stress conditions.

INTRODUCTION

Environmental stresses such as drought and salinity are major factors that limit peanut production in the world [1,2]. It was estimated that drought stress alone is responsible for over $500 million loss in peanut production annually [3]. The major peanut producing countries such as China, India, Nigeria, and USA are all facing severe water shortage for crop irrigation, therefore it is conceivable that the world peanut production will be severely affected in the future. The climate change prediction indicates that more extreme weathers, particularly drought conditions, will occur more often in the tropical and sub-tropical regions of the world [4-6], which will make peanut production extremely challenging in the future. As one of the most nutritious foods for oil and protein source, peanut is very popular among counties in Africa, Asia, and North and South America [7]. Reduced peanut production will affect prices of peanut and peanut derived products such as peanut butter, which will add hardships to people in many countries. Now the great challenge ahead is how to sustain peanut production, and perhaps even increase peanut production for our growing population at the time the condition for peanut production is deteriorating.

Traditional approach in breeding for droughtand salt-tolerant peanut has been slow, due to the rareness of alleles for the drought and salt tolerance in the exiting peanut germplasms [8]. Furthermore, even if a gene for drought or salt tolerance is found in a wild relative of peanut, it will be difficult to introgress the gene into the cultivated peanut cultivars due to the reproductive barrier, not to mention the long time required for many generations of back-crossing. With the advent of molecular biology, biotechnology approach offers a powerful alternative in creating droughtand salt-tolerant peanut in a more efficient way [9-11]. Indeed, over the last 20 years, many genes that confer drought tolerance and salt tolerance were introduced into various crops and tested in laboratories and field, and a few of those genes showed great promises for commercial release [11,12]. For example, the transcription factor genes such as the DREB/ CBF family have been widely used in improving stress tolerance in several crops [13,14]. This class of genes may be useful in creating heatand drought-tolerant peanut in the future (see the example below). In addition to the transcription factor genes, other types of genes might be useful in peanut improvement as well. These genes include those that encode enzymes of ABA biosynthesis or cytokinin biosynthesis pathways, antioxidation metabolisms, and stress signal transduction pathways [8,11]. Some of these functional genes might be useful in improving stress tolerance in peanut in the future.

RECENT PROGRESS IN PEANUT BIOTECHNOLOGY

Peanut Transformation Technologies

To improve peanut's stress tolerance via biotechnology approach, we must be able to transform peanut. Earlier efforts in transforming peanut using Agrobacterium-mediated gene transfer or biolistic bombardment were largely successful [15-23]. However, the transformation efficiencies of these efforts were very low, ranging from 0.3% to less 10%. Sharma and Anjaiah [7] established a peanut transformation procedure that drastically increased peanut transformation efficiency to

55% or higher, which made peanut transformation relatively easy and highly reproducible. More importantly, this new procedure, although still using Agrobacterium as a vehicle, appears to be applicable to many ecotypes of peanut. We were able to use this method to transform both New Mexico Valencia and Texas Runner types at equal frequencies, i.e. ~50% [24-26].

Creation of Drought-Tolerant Peanut

The first report on engineering peanut for increased drought tolerance was from Sharma's group in India [8]. They introduced an Arabidopsis transcription factor gene called AtDREB1A into a drought sensitive peanut line and analyzed transgenic peanut plants for transpiration efficiency under well watered and water-deficit conditions. The transcription factor AtDREB1A binds to the dehydration responsive element (DRE) that is located upstream of many stress responsive genes, thereby activates expression of DRE-containing genes under stressful conditions [27]. Because AtDREB1A was under the control of a stress responsive promoter, Bhatnagar-Mathur et al. [8] were able to obtain transgenic peanut plants that demonstrated similar phenotype to wild-type plants under well watered conditions, yet these transgenic plants showed higher transpiration efficiency under water-deficit conditions. Although no evidence was given in that paper, the AtDREB1A-trasngenic peanut plants are expected to be more drought-, salt-, and low temperature-tolerant, based on known information regarding other AtDREB1A-transgenic plants [28,29].

The second report in literature on genetically engineering peanut for improved drought tolerance was from our effort [24] in which we introduced a cytokinin biosynthetic gene IPT into peanut. The IPT gene encodes isopentenyltransferase, an enzyme that plays a critical role in a rate-limiting step of cytokinin biosynthesis [30]. Drought stress typically causes early senescence in plants, which is advantageous for plants to survive under severe drought conditions in nature as they can quickly finish their life cycle. But this trait is bad for crops, as the yield is severely compromised. If senescence can be delayed under drought conditions, the yield penalty could be reduced. Richmond and Lang [31] demonstrated that exogenous application of cytokinin could delay senescence in plants, and this concept was further proved by later studies that increased cytokinin production by expressing an IPT gene in transgenic plants could indeed delay senescence [32-34]. However, the delayed senescence might not lead to increased seed yield as timing and place of IPT expression in transgenic plants are also important for reducing the yield penalty under drought conditions. Rivero et al. [35] provided an exciting example that regulated expression of IPT by a water-deficit inducible promoter P_{SARK} could significantly reduce the yield penalty caused by drought in transgenic tobacco plants. This strategy worked very well for rice [36], as well as for peanut [24]. The P_{SARK}::IPT-transgenic peanuts plants performed much better than wild-type plants under reduced irrigation conditions in greenhouse, growth chamber, and field conditions. For example, they produced much larger root systems under reduced irrigation in greenhouse conditions, which allowed them to use water more efficiently [24]. They maintained higher photosynthetic rates and stomatal conductance, produced significantly more biomass under reduced irrigation conditions in greenhouse and field conditions (Figures 1(A) and (B)). The yields of P_{SARK}::IPT-transgenic peanuts plants were 30% - 35% higher than that of wild-type plants based on two years of field data [24]. Furthermore, P_{SARK}::IPT-transgenic peanuts plants appeared to produce larger and more multipleseed pods than wild-type peanut plants do (**Figure 2**).

Creation of Salt-Tolerant Peanut

As a glycophytic plant, peanut growth is very sensitive to salt, therefore making peanut salt tolerant will improve peanut production in many parts of the world where soils are saline. In 1999, Blumwald's group demonstrated that overexpression of AtNHX1 that encodes the vacuolar membrane-bound sodium/proton (Na^+/H^+) antiporter in

Figure 1. Wild-type and IPT-transgenic peanut plants in greenhouse (A) and the field (B) under reduced irrigation conditions.

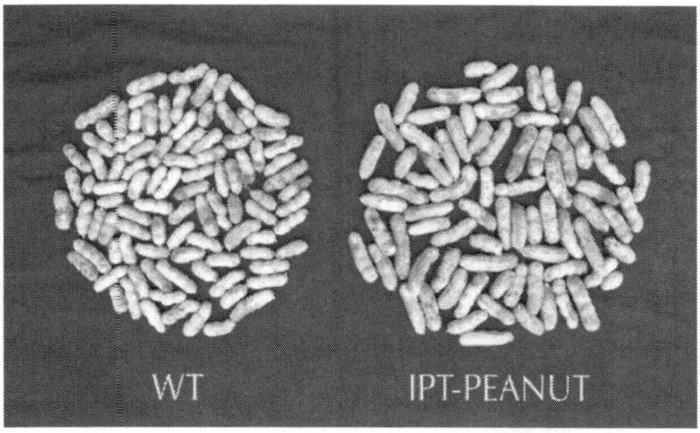

Figure 2. The phenotype of seeds from wild-type and IPTtransgenic peanut plants grown under reduced irrigation conditions.

Arabidopsis could improve salt tolerance in transgenic plants [37]. The increased Na^+/H^+ antiporter activity could lead to increased

Na^+ sequestration into vacuole, which reduces Na^+ toxicity in cytoplasm and at the same time reducing water potential in the vacuole, leading to increased salt tolerance in AtNHX1-overexpressing plants [38]. This approach was successfully used to increase salt tolerance in other plant species such as tomato [39], rapeseed [40], cotton [41], and soybean [42]. We introduced AtNHX1 into peanut and demonstrated that the salt tolerance of peanut could also be improved [26]. AtNHX1-overexpressing peanut plants could tolerate up to 150 mM NaCl in soil, produce higher amounts of biomass, and maintain higher photosynthetic rates and stomatal conductance during salt treatment [26]. Our greenhouse experiments indicated that overexpression of AtNHX1 could confer increased salt tolerance in peanut, however, we have not tested how these AtNHX1-overexpressing peanut plants would behave under field conditions.

Creation of Droughtand Salt-Tolerant Peanut

Recently, we demonstrated that by overexpressing an Arabidopsis vacuolar pyrophosphatase gene AVP1, we could increase drought and salt tolerance simultaneously in transgenic peanut plants [25]. The vacuolar pyrophosphatase is a proton pump that generates proton chemical gradient across vacuolar membranes. Many secondary transporters such as the Na^+/H^+ antipoter (i.e. AtNHX1) depend on the proton chemical gradient generated by proton pump like AVP1 for activities [38,43]. Increasing proton pump activities could increase Na^+/H^+-antipoter activities, thereby increasing salt tolerance [43]. Gaxiola's group demonstrated that overexpression of AVP1 in transgenic plants could increase both droughtand salt-tolerance [44,45]. The increased drought tolerance in the AVP1- overexpresing plants was due to robust root development, which is caused by increased auxin polar transport in transgenic plants [46]. Since the pioneering work by Gaxiola et al. [44], AVP1 and its homologs have been overexpressed in many plants, and in all cases, increased droughtand salt-tolerance have been observed [45,47- 51]. AVP1-overexpressing peanut plants were both droughtand salt-tolerant in greenhouse conditions, as they produced higher amounts of biomass, maintained higher photosynthetic rates and transpiration rates under reduced irrigation and saline conditions [25]. These peanut plants also produced higher yields under reduced irrigation conditions in two-years of field trials [25].

Even though the seed pods from IPT-transgenic peanut appear larger than those from wild-type peanut plants (**Figure 2**), there were no major changes in the oil contents between IPT-transgenic peanut seeds and wild-type peanut seeds [24]. Similarly, there were no major changes in the oil contents between AVP1-transgenic and wildtype peanut seeds as well [25]. The contents of major fatty acids such as palmitic acid ($C_{16:0}$), oleic acid ($C_{18:1}$) and linoleic acid ($C_{18:2}$) were similar between wild-type and IPT-transgenic and AVP1-transgenic peanut plants. The minor fatty acids varied a little between transgenic peanut plants and wild-type plants, but these differences were also found between wild-type and the segregated non-transgenic lines [24,25], suggesting that they might be due to somatic variation resulting from tissue cultures. Furthermore, these differences are likely found among different peanut ecotypes, which show little nutritional differences. It is apparent that overexpression of AVP1 and regulated expression of IPT in peanut do not affect the oil content and major fatty acid compositions. Whether or not P_{SARK}::IPT-transgenic peanut and AVP1-expressing peanut plants would increase peanut yield under field conditions in large-scale trials is not known, but the data we have from laboratory experiments as well as small-scale field experiments are very promising, which strongly suggest that IPT and AVP1 are excellent target genes that could improve peanut production in areas where water is limited.

FUTURE CHALLENGES

Improving Peanut Transformation Efficiency

Although the procedure developed by Sharma and Anjaiah [7] is quite good in transforming peanut, the false positive rate of over 40% among the regenerated peanut lines is still too high, as sorting out false positives is time-consuming and resource-wasting. Kanamycin was used as the selection marker in transformation, but peanut appears to have some resistance to kanamycin, which leads to a large numbers of escapers in the regeneration process. Because of the resistance to kanamycin, transgenic peanut cannot be screened directly at seedling stage in kanamycin media, instead transgenic peanut plants have to be screened by DNA blot, PCR or RNA blot analysis, which makes characterization of transgenic peanut plants labor-intensive and time-consuming. Therefore it is necessary to develop a more efficient system for peanut transformation, perhaps a better antibiotics that is more effective in killing non-transgenic peanut plants needs to be identified as the selective marker.

Finding More Efficient Genes for Peanut Improvement

The effect of expressing AtDREB1A in peanut under the control of a stress responsive promoter on peanut yield in field conditions has yet to be reported, it is difficult to estimate the impact of this gene on improving peanut production in the semi-arid regions of the world. Based on our own experiences on working with IPT-transgenic peanut and AVP1-transgenic peanut, we have the following observations: IPT-transgenic peanut plants consistently produced higher yields than AVP1-trasngenic plants, and both consistently outperformed wild-type peanut plants (on averages 30% and 20% better than wild-type plants, respectively) under reduced irrigation conditions in the semi-arid regions of West Texas [24,25]. However, IPTtransgenic peanut plants are not salt-tolerant, yet AVP1- transgenic peanut plants are. Consequently, each has its own strength and weakness. Other genes that prove to be effective in conferring drought tolerance in other transgenic crops have not been tested in peanut, their potential use in peanut might be worth of trying. For example, the SNAC1 gene from rice was shown to confer 23% - 34% higher yields when overexpressed in transgenic rice plants under drought conditions [52]. The yield differences between SNAC1-transgenic rice and wild-type rice are similar to the differences found between IPT-transgenic peanut and wild-type peanut plants under water-deficit conditions. Recently a gene named OsSIZ1 was shown to significantly increase both droughtand heat-tolerance when overexpressed in transgenic creeping bentgrass [53], and similar observation was made in transgenic cotton plants expressing the OsSIZ1 transgene (our unpublished data), suggesting that OsSIZ1 would be an excellent gene that could be used for improving peanut. There might be more genes out there that confer similar or higher tolerance to drought stress, we need to find an efficient way to test them in transgenic peanut plants in the future.

GMO Issues

Perhaps the biggest challenge that we face today in plant biotechnology is not a technical issue, but a political or philosophic issue: should genetically modified organisms (GMO) or genetic engineering products such as transgenic peanut plants be developed and marketed? One of the major obstacles for the adoption of GMO in Europe and many other countries is the perception that GMO is not safe. Although the safety issue has been carefully addressed for each GMO product that is approved for commercial release, there is still strong opposition against GMO products. We have to admit that we did not do a good job in educating general public about the nature of changes in each GMO product, which led to misunderstanding of the GMO products by the general public. There are a number of GMO products on market and these GMO products contain either herbicide tolerance gene or insect resistance gene, which were inserted into crops either as a single trait or stacked traits [54]. The benefits of having crops with these genes far outweigh the potential harm that these crops might bring, in fact after nearly 20 years of growing crops with these traits and more than 134 million hectares of transgenic crops grown by 2009 [55], there was not a single report on the adverse impact of GMO on food safety or environmental damage [56]. In US, three governmental agencies oversee the testing and approval of genetically engineered products: the US Department of Agriculture (USDA), Food and Drug Administration (FDA), and Environmental Protection Agency (EPA) [56]. USDA has strict regulations for testing all GMOs before their release, and is responsible for protecting US agriculture and environment from potential pests. FDA is responsible for ensuring the safety of GMO as human's food or animal feed, and EPA is responsible for ensuring that genetically engineered pesticides pose no threats to human and our environment [56]. All GMO products on the market had been thoroughly tested and are continuingly to be carefully scrutinized. If any potential harm to human health or to our environment is found, they will be removed from market immediately. The benefits of using GMOs have been enormous: reduction of 17% pesticides usage from 1996 to 2007 in the US alone (about 359,000 tons of chemicals were not applied in the US soil) and over $19 billion more revenue generated for US farmers during this period [54].

Another misperception about GMO is that man should not change the genetic make-up of crops, as crop plants have evolved for so long to reach today's state, there must be a reason why crops should be the way they are. People who think this way do not realize that in fact there were no crops that would look like what they are today even just 100 years ago. All crop plants we have today were derived from many years' selection and breeding by people, the crops' genomes had changed dramatically over the last century. The yields of crops that existed several thousand years ago were extremely low, usually less than 5% of today's yields. For example, corn's wild relatives today produce ears that are about 2 cm long with tiny kernels. Furthermore, the wild relatives of corn or rice not only have low yields, but low quality as well. They usually taste bad, with low nutritional values, and many even have high levels of allergens. Over the years farmers had selected the best varieties to breed, and especially since the discovery of Mendelian's rules of genetics in early 1910s, scientists had bred crops with the best agronomic traits and created modern varieties of today's crops. People never questioned crops created by traditional breeding methods in the last 100 years, but in fact traditional breeding involves manipulating hundreds, perhaps thousands of genes during crossing between different parental lines or lines from exotic germplasms. Whereas, GMO products usually involve one or a few genes whose biochemistry has been studied extensively, and whose potential being allergens or toxicants is very small, if existent [56]. In summary, given the strict regulations from government and thorough studies from academia, the GMO resulted from biotechnology should be safer, not the other way around [57,58]. We need to pass this information to general public as one of our missions, if we want to continue to succeed in creating better and safer biotechnology products in the future.

ACKNOWLEDGEMENTS

Our research was supported by grants from Texas Peanut Producers Board, US National Peanut Board, and USDAOgallala Aquifer Program.

REFERENCES

1. J. R. Stansell and J. E. Pallas, "Yield and Quality Response of Florunner Peanut to Applied Drought at Several Growth Stages," Peanut Science, Vol. 12, No. 2, 1985, pp. 64-70.doi:10.3146/pnut.12.2.0005
2. M. C. Lamb, J. I. Davidson, J. W. Childre and N. R. Martin, "Comparison of Peanut Yield, Quality, and Net Returns between Nonirrigated and Irrigated Production," Peanut Science, Vol. 24, No. 2, 1997, pp. 97-101. doi:10.3146/i0095-3679-24-2-7
3. K. K. Sharma and M. Lavanya, "Recent Developments in Transgenics for Abiotic Stress in Legumes of the SemiArid Tropics," In: M. Ivanaga, Ed., Genetic Engineering of Crop Plants for Abiotic Stress, JIRCAS Working Report No. 23, JIRCAS, Tsukuba, 2002, pp. 61-73.
4. D. B. Lobell, M. B. Burke, C. Tebaldi, M. D. Mastrandrea, W. P. Falcon and R. L. Naylor, "Prioritizing Climate Change Adaptation Needs for Food Security in 2030," Science, Vol. 319, No. 5863, 2008, pp. 607-610. doi:10.1126/science.1152339
5. D. S. Battisti and R. L. Naylor, "Historical Warnings of Future Food Insecurity with Unprecedented Seasonal Heat," Science, Vol. 323, No. 5911, 2009, pp. 240-244.doi:10.1126/science.1164363
6. S. P. Long and D. R. Ort, "More than Taking the Heat: Crops and Global Change," Current Opinion in Plant Biology, Vol. 13, No. 3, 2010, pp. 241-248.doi:10.1016/j.pbi.2010.04.008
7. K. Sharma and V. Anjaiah, "An Efficient Method for the Production of Transgenic Plants of Peanut (Arachis hypogaea L.) through Agrobacterium tumefaciens-Mediated genetic Transformation," Plant Science, Vol. 159, No. 1, 2000, pp. 7-19. doi:10.1016/S0168-9452(00)00294-6
8. P. Bhatnagar-Mathur, et al., "Stress-Inducible Expression of AtDREB1A in Transgenic Peanut (Arachis hypogaea L.) Increases Transpiration Efficiency under Water-Limiting Conditions," Plant Cell Reports, Vol. 26, No. 12, 2007, pp. 2071-2082.doi:10.1007/s00299-007-0406-8
9. R. Mittler and E. Blumwald, "Genetic Engineering for Modern Agriculture: Challenges and Perspectives," Annual Review of

Plant Biology, Vol. 61, 2010, pp. 443-462.doi:10.1146/annurev-arplant-042809-112116

10. R. K. Varshney, K. C. Bansal, P. K. Aggarwal, S. K. Datta and P. Q. Craufurd, "Agricultural Biotechnology for Crop Improvement in a Variable Climate: Hope or Hype?" Trends in Plant Science, Vol. 16, No. 7, 2011, pp. 363-371. doi:10.1016/j.tplants.2011.03.004
11. J. Deikman, M. Petracek and J. E. Heard, "Drought Tolerance through Biotechnology: Improving Translation from the Laboratory to Farmers' Fields," Current Opinion in Biotechnology, Vol. 23, No. 2, 2012, pp. 243-250. doi:10.1016/j.copbio.2011.11.003
12. S. Yang, B. Vanderbeld, J. Wang and Y. Huang, "Narrowing Down the Targets: Towards Successful Genetic Engineering of Drought-Tolerant Crops," Molecular Plant, Vol. 3, No. 3, 2010, pp. 469-490. doi:10.1093/mp/ssq016
13. K. Century, T. L. Reuber and O. J. Ratcliffe, "Regulating the Regulators: The Future Prospects for TranscriptionFactor-Based Agricultural Biotechnology Products," Plant Physiology, Vol. 147, No. 1, 2008, pp. 20-29. doi:10.1104/pp.108.117887
14. D. Todaka, N. Kazuo, S. Kazuo and Y.-S. Kazuko, "Toward Understanding Transcriptional Regulatory Networks in Abiotic Stress Responses and Tolerance in Rice," Abiotic Stress in Rice, Vol. 5, 2012, p. 6. http://www.thericejournal.com/content/5/1/6
15. P. Ozias-Akins, et al., "Regeneration of Transgenic Peanut Plants from Stably Transformed Embryogenic Callus," Plant Science, Vol. 93, No. 1-2, 1993, pp. 185-194.doi:10.1016/0168-9452(93)90048-5
16. M. Cheng, D. C. H. His and G. C. Phillips, "Recovery of Transformants of Valencia-Type Peanut Using Agrobacterium tumefaciens," Peanut Science, Vol. 21, No. 2, 1994, pp. 84-88. doi:10.3146/i0095-3679-21-2-3
17. M. Cheng, R. L. Jarret, Z. Li, A. Xing and J. W. Demski, "Production of Fertile Transgenic Peanut (Arachis hypogaea L.) Plants Using Agrobacterium tumefaciens," Plant Cell Reports, Vol. 15, No. 9, 1996, pp. 653-657. doi:10.1007/BF00231918
18. S. Eapen and L. George, "Agrobacterium tumefaciens Mediated Gene Transfer in Peanut (Arachis hypogaea L.)," Plant Cell Reports, Vol. 13, No. 10, 1994, pp. 582-586.doi:10.1007/BF00234516

19. D. M. Livingstone and R. G. Birch, "Plant Regeneration and Microprojectile-Mediated Gene Transfer in Embryonic Leaflets of Peanut (Arachis hypogaea L.)," Australian Journal of Plant Physiology, Vol. 22, No. 4, 1995, pp. 585-591. doi:10.1071/PP9950585

20. C. Singsit, et al., "Expression of a Bacillus Thuringiensis cryIA(c) Gene in Transgenic Peanut Plants and Its Efficiency against Lesser Cornstalk Borer," Transgenic Research, Vol. 6, No. 2, 1997, pp. 169-176. doi:10.1023/A:1018481805928

21. H. Yang, C. Singsit, A. Wang, D. Gonsalves and P. OziasAkins, "Transgenic Peanut Plants Containing a Nucleocapsid Protein Gene of Tomato Spotted Wilt Virus Show Divergent Levels of Gene Expression," Plant Cell Reports, Vol. 17, No. 9, 1998, pp. 693-699.doi:10.1007/s002990050467

22. A. H. McKently, G. A. Moore, H. Doostdar and R. P. Niedz, "Agrobacterium-Mediated Transformation of Peanut (Arachis hypogaea L.) Embryo Axes and the Development of Transgenic Plants," Plant Cell Reports, Vol. 14, No. 11, 1995, pp. 699-703. doi:10.1007/BF00232650

23. V. K. Rohini and K. S. Rao, "Transformation of Peanut (Arachis hypogaea L.): A Non-Tissue Culture Based Approach for Generating Transgenic Plants," Plant Science, Vol. 150, No. 1, 2000, pp. 41-49. doi:10.1016/S0168-9452(99)00160-0

24. H. Qin, et al., "Regulated Expression of an Isopentenyltransferase Gene (IPT) in Peanut Significantly Improves Drought Tolerance and Increases Yield under Field Conditions," Plant and Cell Physiology, Vol. 52, No. 11, 2011, pp. 1904-1914. doi:10.1093/pcp/pcr125

25. H. Qin, et al., "Expression of the Arabidopsis Vacuolar H^+-Pyrophosphatase Gene AVP1 in Peanut to Improve Drought and Salt Tolerance," Plant Biotechnology Reports, 2012.doi:10.1007/s11816-012-0269-5

26. M. Banjara, L. Zhu, G. Shen, P. Payton and H. Zhang, "Expression of an Arabidopsis Sodium/Proton Antiporter Gene (AtNHX1) in Peanut to Improve Salt Tolerance," Plant Biotechnology Report, Vol. 6, 2011, pp. 59-67.

27. K. Yamaguchi-Shinozaki and K. Shinozaki, "Organizantion of Cis-Acting Regulatory Elements in Osmoticand Cold-Stress-

Responsive Promoters," Trends in Plant Science, Vol. 10, 2005, pp. 88-94. doi:10.1016/j.tplants.2004.12.012

28. Y. Ito, et al., "Functional Analysis of Rice DREB1/CBFType Transcription Factors Involved in Coldresponsive Gene Expression in Transgenic Rice," Plant Biotechnology Reports, Vol. 47, No. 1, 2006, pp. 141-153. doi:10.1093/pcp/pci230
29. K. Datta, N. Baisakh, M. Ganguly, S. Krishnan, K. Yamaguchi-Shinozaki and S. K. Datta, "Overexpression of Arabidopsis and Rice Stress Genes' Inducible Transcription Factor Confers Drought and Salinity Tolerance to Rice," Plant Biotechnology Journal, Vol. 10, No. 5, 2012, pp. 579-586. doi:10.1111/j.1467-7652.2012.00688.x
30. H. Sakakibara, "CYTOKININS: Activity, Biosynthesis, and Translocation," Annual Review of Plant Biology, Vol. 57, 2006, pp. 431-449.doi:10.1146/annurev.arplant.57.032905.105231
31. A. E. Richmond and A. Lang, "Effect of Kinetin on Protein Content and Survival of Detached Xanthium Leaves," Science, Vol. 125, No. 3249, 1957, pp. 650-651.doi:10.1126/science.125.3249.650-a
32. C. M. Smart, S. R. Scofield, M. W. Bevan and T. A. Dyer, "Delayed Leaf Senescence in Tobacco Plants Transformed with tmr, a Gene for Cytokinin Production in Agrobacterium," Plant Cell, Vol. 3, 1991, pp. 647-656.
33. B. Martineau, C. M. Houck, R. E. Sheehy and W. R. Hiatt, "Fruit-Specific Expression of the A. tumefaciens Isopentenyl Transferase Gene in Tomato: Effects on Fruit Ripening and Defense-Related Gene Expression in Leaves," Plant Journal, Vol. 5, No. 1, 1994, pp. 11-19. doi:10.1046/j.1365-313X.1994.5010011.x
34. S. Gan and R. M. Amasino, "Inhibition of Leaf Senescence by Autoregulated Production of Cytokinin," Science, Vol. 270, No. 5244, 1995, pp. 1986-1988.doi:10.1126/science.270.5244.1986
35. R. M. Rivero, et al., "Delayed Leaf Senescence Induces Extreme Drought Tolerance in a Flowering Plant," Proceedings of National Academy of Sciences of the United States of America, Vol. 104, No. 49, 2007, pp. 19631- 19636. doi:10.1073/pnas.0709453104
36. Z. Peleg, M. Reguera, H. Walia and E. Blumwald, "Cytokinin Mediated Source-Sink Modifications Improve Drought Tolerance and Increases Grain Yield in Rice under Water Stress," Plant

Biotechnology Journal, Vol. 9, No. 7, 2011, pp. 747-758. doi:10.1111/j.1467-7652.2010.00584.x

37. M. P. Apse, G. S. Aharon, W. A. Snedden and E. Blumwald, "Salt Tolerance Conferred by Overexpression of a Vacuolar Na^+/ H^+ Antiport in Arabidopsis," Science, Vol. 285, No. 5431, 1999, pp. 1256-1258. doi:10.1126/science.285.5431.1256
38. E. Blumwald, "Salt Transport and Salt Resistance in Plants and Other Organisms," Current Opinion in Cell Biology, Vol. 12, No. 4, 2000, pp. 431-434. doi:10.1016/S0955-0674(00)00112-5
39. H. X. Zhang and E. Blumwald, "Transgenic Salt-Tolerant Tomato Plants Accumulate Salt in Foliage but Not in Fruit," Nature Biotechnology, Vol. 19, 2001, pp. 765-768.doi:10.1038/90824
40. H. X. Zhang, J. N. Hodson, J. P. Williams and E. Blumwald, "Engineering Salt-Tolerant Brassica Plants: Characterization of Yield and Seed Oil Quality in Transgenic Plants with Increased Vacuolar Sodium Accumulation," Proceedings of National Academy of Sciences of the United States of America, Vol. 98, No. 22, 2001, pp. 12832- 12836.doi:10.1073/pnas.231476498
41. C. He, et al., "Expression of an Arabidopsis Vacuolar Sodium/ Proton Antiporter Gene in Cotton Improves Photosynthetic Performance under Salt Conditions and Increases Fiber Yield in the Field," Plant Biotechnology Reports, Vol. 46, No. 11, 2005, pp. 1848-1854.doi:10.1093/pcp/pci201
42. T. X. Li, et al., "Stable Expression of Arabidopsis Vacuolar Na^+/ H^+ Antiporter Gene AtNHX1, and Salt Tolerance in Transgenic Soybean for over Six Generations," Chinese Science Bulletin., Vol. 55, No. 12, 2010, pp. 1127- 1134. doi:10.1007/s11434-010-0092-8
43. R. A. Gaxiola, G. R. Fink and K. D. Hirschi, "Genetic Manipulation of Vacuolar Proton Pumps and Transporters," Plant Physiology, Vol. 129, No. 3, 2002, pp. 967- 973.doi:10.1104/pp.020009
44. R. A. Gaxiola, et al., "Droughtand Salt-Tolerant Plants Result from Overexpression of the AVP1 H^+-Pump," Proceedings of National Academy of Sciences of the United States of America, Vol. 98, No. 20, 2001, pp. 11444- 11449. doi:10.1073/pnas.191389398
45. S. Park, et al., "Up-Regulation of a H^+-Pyrophosphatase (H^+-PPase) as a Strategy to Engineer Drought-Resistant Crop Plants,"

Proceedings of National Academy of Sciences of the United States of America, Vol. 102, No. 52, 2005, pp. 18830-18835. doi:10.1073/pnas.0509512102

46. J. Li, et al., "Arabidopsis H^+-PPase AVP1 Regulates AuxinMediated Organ Development," Science, Vol. 310, No. 5745, 2005, pp. 121-125. doi:10.1126/science.1115711
47. F. Y. Zhao, X. J. Zhang, P. H. Li, Y. X. Zhao and H. Zhang, "Co-Expression of the Suaeda salsa SsNHX1 and Arabidopsis AVP1 Confer Greater Salt Tolerance to Transgenic Rice than the Single SsNHX1," Molecular Breeding, Vol. 17, No. 4, 2006, pp. 341-353.doi:10.1007/s11032-006-9005-6
48. B. Li, A. Wei, C. Song, N. Li and J. Zhang, "Heterologous Expression of the TsVP Gene Improves the Drought Resistance of Maize," Plant Biotechnology Journal, Vol. 6, No. 2, 2008, pp. 146-159. doi:10.1111/j.1467-7652.2007.00301.x
49. S. Lv, K. Zhang, Q. Gao, L. Lian, Y. Song and J.-R. Zhang, "Overexpression of an H^+-PPase from Thellungiella halophila in Cotton Enhances Salt Tolerance and Improves Growth and Photosynthetic Performance," Plant Biotechnology Reports, Vol. 49, No. 8, 2008, pp. 1150-1164. doi:10.1093/pcp/pcn090
50. S. Lv, L.-J. Lian, P. L. Tao, Z.-X. Li, K.-W. Zhang and J.-R. Zhang, "Overexpression of Thellungiella halophila H^+-PPase (TsVP) in Cotton Enhances Drought Stress Resistance of Plants," Planta, Vol. 229, No. 4, 2009, pp. 899-910. doi:10.1007/s00425-008-0880-4
51. V. Pasapula, et al., "Expression of an Arabidopsis Vacuolar H^+-Pyrophosphatase Gene (AVP1) in Cotton Improves Droughtand Salt-Tolerance and Increases Fiber Yield in the Field Conditions," Plant Biotechnology Journal, Vol. 9, No. 1, 2011, pp. 88-99. doi:10.1111/j.1467-7652.2010.00535.x
52. H. Hu, et al., "Overexpressing a NAM, ATAF, and CUC (NAC) Transcription Factor Enhances Drought Resistance and Salt Tolerance in Rice," Proceedings of National Academy of Sciences of the United States of America, Vol. 103, No. 35, 2006, pp. 12987-12992. doi:10.1073/pnas.0604882103
53. Z. Li, et al., "Heterologous Expression of OsSIZ1, a Rice SUMO E3 Ligase, Enhances Broad Abiotic Stress Tolerance in Transgenic

Creeping Bentgrass," Plant Biotechnology Journal, Vol. 11, No. 4, 2012, pp. 432-445. doi:10.1111/pbi.12030

54. J. R. Park, I. McFarlane, R. H. Phipps and G. Ceddia, "The Role of Transgenic Crops in Sustainable Development," Plant Biotechnology Journal, Vol. 9, No. 1, 2011, pp. 2-21.doi:10.1111/j.1467-7652.2010.00565.x
55. C. James, "Global Status of Commercialised Biotech/GM Crops," ISAAA Brief 41, Executive Summary, 2009. http://www.isaaa.org/resources/publications/briefs/41/executivesummary/default.asp
56. A. McHughen and S. Smyth, "US Regulatory System for Genetically Modified [Genetically Modified Organism (GMO), rDNA or Transgenic] Crop Cultivars," Plant Biotechnology Journal, Vol. 6, No. 1, 2008, pp. 2-12.
57. P. G. Lemaux, "Genetically Engineered Plants and Foods: A Scientist's Analysis of the Issues (Part I)," Annual Review of Plant Biology, Vol. 59, 2008, pp. 771-812.doi:10.1146/annurev.arplant.58.032806.103840
58. P. G. Lemaux, "Genetically Engineered Plants and Foods: A Scientist's Analysis of the Issues (Part II)," Annual Review of Plant Biology, Vol. 60, 2009, pp. 511-559.doi:10.1146/annurev.arplant.043008.092013

Chapter 3

Impact of Molecular Genetic Research on Peanut Cultivar Development

C. Corley Holbrook [1], Peggy Ozias-Akins [2], Ye Chu [2] and Baozhu Guo [3]

[1]USDA-ARS, Crop Genetics and Breeding Research Unit, 115 Coastal Way, Tifton, GA 31793, USA

[2]Department of Horticulture and NESPAL, University of Georgia, 2360 Rainwater Rd, Tifton, GA 31794, USA;

[3]USDA-ARS, Crop Protection and Management Research Unit, 2747 Davis Rd, Tifton, GA 31793, USA;

ABSTRACT

Peanut (*Arachis hypogaea* L.) has lagged other crops on use of molecular genetic technology for cultivar development in part due to lack of investment, but also because of low levels of molecular polymorphism among cultivated varieties. Recent advances in molecular genetic technology have allowed researchers to more precisely measure genetic polymorphism and enabled the development of low density genetic maps for *A. hypogaea* and the identification of molecular marker or QTL's for several economically significant traits. Genomic research has also been used to enhance the amount of genetic diversity available for use in conventional breeding through the development of transgenic peanut, and the creation of TILLING populations and synthetic allotetraploids. Marker assisted selection (MAS) is becoming more common in peanut cultivar development programs, and several cultivar releases are anticipated in the near future. There are also plans to sequence the peanut genome in the near future which should result in the development of additional molecular tools that will greatly advance peanut cultivar development.

DEVELOPMENT OF TOOLS TO ENHANCE MOLECULAR BREEDING IN PEANUT

Genomic research can provide new tools and resources to revolutionize crop genetic improvement and production [1]. However, genomic research in peanut (*Arachis hypogaea* L.) is far behind that in other crops such as maize, soybean, wheat, sorghum, and potato due to the shortage of essential genome infrastructure, tools, and resources [2]. As a consequence, peanut has lagged behind other crops on the use of molecular genetic technology for cultivar development. The early technologies (isozyme, RFLP (Restriction Fragment Length Polymorphism), AFLP (Amplified Fragment Length Polymorphism), RAPD (Random Amplified Polymorphic DNA), and SCAR (Sequence Characterized Amplified Region)) showed extremely low levels of polymorphism in *A. hypogaea* [3-11]. Those early struggles have been documented in several excellent reviews [2,12-14]. Recent advances in molecular genetic technology have allowed researchers to detect more frequent genetic polymorphism. These efforts have resulted in the construction of moderate density genetic maps for *A. hypogaea* [15-19] populated primarily with SSR (simple sequence repeat or microsatellite) markers that contrast with other PCR-based markers in their largely co-dominant *vs.* dominant (AFLP, RAPD, and SCAR) nature. Many of these SSR markers were developed from peanut ESTs (expressed sequence tags). Because of genome size and complexity, many plant EST libraries have been sequenced as an alternative to whole genome sequences, including peanut. EST data sets were foundational for functional genomics during the period when only a few plant genomes were sequenced and before the development of the second generation of high throughput sequencing technology. ESTs have been especially important resources for major crops or economically significant plants with large genomes (such as peanut) to enable gene discovery, gene expression analysis and molecular marker development.

The NCBI EST database contains 225,264 ESTs from peanut as of November 2011 [20]. There are 150,922 for *A. hypogaea* (including 745 for subsp. *fastigiata*), 35,291 for *A. duranensis*, 32,787 for *A. ipaensis*, and 6264 for *A. stenosperma*. Many of the *A. hypogaea* ESTs have been combined with short-read sequences to create a first generation transcriptome assembly (NCBI BioProject PRJNA49471). Before the completion of peanut whole genome sequence, sequencing large numbers of ESTs can create a formidable resource for studies in both biodiversity and gene-discovery. Sequence analysis tools have extended the scope of EST utility into the fields of proteomics, marker development and genome annotation. Although EST collections certainly are not intended to substitute for a whole genome sequence, the EST resource forms the core foundation for various genome-wide experiments, particularly for microarray gene expression study [21,22], marker development and genetic map construction [17], which will assist assembly of the whole genome. The ESTs will continue to be actively sequenced to fill knowledge gaps and complement the whole genome sequence.

In spite of the discovery of thousands of microsatellite-containing EST and genomic sequences from which markers have been developed [23], only ~10–20% detect multiple alleles among tetraploid peanut genotypes [24-27], although somewhat higher levels of polymorphism have been observed in several studies [28-32]. In spite of considerable molecular tool expansion for *A. hypogaea* over the past decade, low polymorphism resulting from a genetic bottleneck due to polyploidization [33] continues to limit the number of markers that can be mapped in populations from intraspecific biparental crosses. The discovery of SNP (Single Nucleotide Polymorphism) markers will further enhance the molecular toolkit for peanut, although high-throughput SNP genotyping in the tetraploid will be challenging [23].

From a cultivar development standpoint, however, these advances in technology have enabled the identification of molecular markers associated with quantitative trait loci (QTLs) for several economically significant traits. Recent research has resulted in the discovery of molecular markers associated with resistance to foliar diseases, rust and late leaf spot [18,34-36], resistance to *Cylindrocladium* black rot and early leaf spot [14], nematode resistance [37-40], resistance to TSWV [17], resistance to the aphid vector of groundnut rosette disease [41], drought tolerance [15,42], yield parameters [43], high oleic acid [44,45], and seed biochemical traits [46]. Many of these QTLs are not major, *i.e.*, they account for <10% of the phenotypic variation explained. Major QTLs identified for rust and late leaf spot in at least one germplasm source may be of wild species origin [18,36,47] as is the source of nematode resistance [40,48]. Many small effect QTLs were mapped in a population segregating for drought tolerance and its surrogate traits such as transpiration efficiency, specific leaf area, or dry weight and the percent of phenotypic variation explained was often low (<10%) [15,42]. Two additional populations confirmed that no major QTLs for drought tolerance could be identified [19]. The need to pyramid a large number of minor QTLs for drought tolerance may predetermine the most efficient breeding strategy to integrate with marker-assisted selection [49]. For example, marker-assisted backcrossing can efficiently combine only a few genes for foreground (trait-associated donor alleles) selection while conducting background selection using markers spanning the genome in order to rapidly recover the recurrent parent genotype plus the genes/alleles of interest. Foreground selection becomes more costly as the number of minor QTLs increases because population sizes increase dramatically. Up to now, neither background nor genome-wide selection has been practiced in peanut and must await the development of high-throughput, economical assays for large numbers of markers. For other traits such as high oleic acid or rust resistance, identification of major QTLs have or will enable efficient marker-assisted backcrossing [50].

CREATING NEW SOURCES OF GENETIC DIVERSITY IN PEANUT

Molecular breeding in peanut trails that of many crops [51,52] in part due to a lack of investment, but also because of low levels of molecular polymorphism among cultivated varieties. While polymorphism is abundant in diploid wild species, it is sparse in tetraploid peanut due to the genetic bottleneck imposed by its relatively recent origin [33]. Genomic research might also be used to enhance the amount of genetic diversity available for application in conventional breeding. The development of transgenic peanut, and the creation of TILLING populations and synthetic allotetraploids are three approaches that are being explored to create new sources of genetic diversity in peanut.

Transgenics

Transgenic research has resulted in the creation of new genetic diversity in peanut which could be very useful in cultivar development. The first successful transformation of peanut was achieved using the biolistic/bombardment technique with accompanying plant regeneration [53]. Since then several research groups have successfully employed this biolistic method using embryogenic cultures as the target tissues, and a few groups have utilized *Agrobacterium*-mediated transformation which relies primarily on shoot-regenerating cultures. The biolistic method directly transfers target genes into plant cells by delivering DNA coated microprojectiles at a high velocity [54]. The direct delivery of transgenes makes the transformation less dependent on host genotype. As long as a peanut genotype can be regenerated from somatic embryo tissue culture, it can be transformed by bombardment. Three major peanut cultivar groups including runner, spanish and virginia have been transformed by this method. Besides the target gene of interest, DNA constructs for bombardment often include a selectable marker gene cassette expressing hygromycin phosphotransferase (*hph* or *hpt*). After bombardment, embryogenic tissues are cultured on selective medium to minimize non-transgenic escapes [53] and to yield highly proliferating transformed tissues that are further transferred to regeneration medium for shoot and root induction. Genetic transformation via bombardment has successfully transferred multiple genes for protein accumulation and

RNAi-mediated gene silencing. Limitations for biolistic bombardment include: (1) it is a lengthy process which takes 12–14 months from initiation of tissue culture to the establishment of primary transgenic plants; and (2) infertility is frequent among tissue culture regenerants [53,55]. On the other hand, *Agrobacterium tumefaciens*-mediated transformation circumvents the tissue culture step and takes less time (usually 4–5 months) to obtain transgenic plants, but it is highly genotype dependent. In this case, a target gene of interest is engineered into the T-DNA region of a 'disarmed' plasmid and introduced into *A. tumefaciens*. The transgene within the T-DNA borders is further transferred into plant cells by cocultivation of *A. tumefaciens* and wounded plant tissue. Induction of *Agrobacterium* virulence genes is critical for transgene integration into the plant cell genome and is affected by exposure to phenolic compounds (such as acetosyringone), reducing monosaccharides and acidic pH by host cells [56,57].

Various peanut tissues including leaf sections, cotyledonary nodes, longitudinal cotyledon halves, embryo axes, embryo leaflets, and hypocotyls have been tested for *A. tumefaciens* transformation [58-60]. Apical or axillary meristematic cells in these tissues allow for multiple shoot regeneration and have been targeted for gene transfer by *A. tumefaciens*. However, conditions for adventitious shoot formation through organogenesis vary widely, and cocultivation protocols with or without virulence inducing agents have been previously reviewed [61,62]. Complex host-pathogen interaction limits peanut genotypes that can be successfully transformed by *Agrobacterium*. Out of 19 publications on peanut transformation mediated by *A. tumefaciens*, 14 of them performed transformation mainly with either spanish (such as JL-24 and TMV 2) or valencia (such as New Mexico Valencia A) types of peanut. Runner cultivars account for 80% of the production in the U.S. Only one publication reported transgenic runner type peanut by *A. tumefaciens*-mediated transformation [63]. Differential expression of host genes in the first few hours of *A. tumefaciens* cocultivation can affect T-DNA integration and transformation efficiency as demonstrated in rice [64]. Recently, direct shoot organogenesis was achieved with a couple of US runner type cultivars Georgia Green [65] and Florida-07 [66] at a shoot production rate of 25% and 7% respectively [67], a frequency considerably lower than the 82–90% reported for the Indian cultivar JL-24 [59,68]. Further transformation study would potentially implement this transformation technology into major US peanut cultivars.

Specific details of peanut transformation events up to year 2005 have been documented in several reviews [62,69,70]. Peanut transgenic research since year 2006 is summarized in Table 1. Besides a couple of studies on transformation efficiency and selection conditions [68,71], more recent advances in peanut transformation mainly focus on integrating genes conferring resistance to biotic and abiotic stresses. To improve peanut drought tolerance, *AtDREB1A*, a cis-acting transcription factor that binds to dehydration responsive element (DRE) from*Arabidopsis thaliana,* was transformed to peanut under the control of a stress inducible promoter from the rd29A gene [72]. One transgenic line demonstrated a 40% increase in transpiration efficiency (TE) in a greenhouse drought tolerance test. Further analysis of antioxidative responses from these transgenic lines cannot provide an explanation for the elevated TE performance [73]. In addition, improved greenhouse drought and salt tolerance was found among transgenic peanut lines transformed with *AtNHXI*, a vacuolar Na^+/H^+ antiporter [74]. Isopentenyltransferase (*IPT*), a key enzyme in the cytokinin biosynthesis pathway, driven by a drought inducible SARK promoter was used to transform peanut [75]. Transgenic lines demonstrated improved biomass retention in a greenhouse drought tolerance test and an average of 58% yield increase in a two-year field test. Transgenic peanut expressing a human Bcl-xL gene has improved tolerance to paraquat, a bipyridilium herbicide [76]. Fungal resistance in peanut was enhanced by transforming several genes including barley oxlate oxidase [77,78], mustard defensin [60], rice chitinase [79,80] and chloroperoxidase [81]. Evaluation of some transgenic lines was advanced to field studies such as resistance to *Sclerotinia minor*, which was confirmed in oxlate oxidase and rice chitinase transformed lines [78,79]. Synthetic *cry*1 EC transformed peanut was shown to confer resistance to the larvae of *Spodoptera litura* [82].

Table 1. Peanut Genetic transformation since year 2006

Peanut Genotype	Transformation Method	Explant	Promoter	Transgene	Selectable Marker	Trait Evaluation	Reference
JL-24	*A. tumefaciens* strain GV2260	cotyledonary node	CaMV 35 S	GUS	*npt*II	none	68
JL-24	*A. tumefaciens* strain C58	cotyledon	*A. thaliana* rd29A	*AtDREB1A*	*npt*II	green house tests for drought tolerance antioxidative response to drought stress	72 73
JL-24	*A. tumefaciens* strain EHA105	embyro axes	CaMV 35 S	mustard defensin	*npt*II	detached leaf assay and greenhouse tests for late leaf spot resistance	60
JL-24	*A. tumefaciens* strain EHA101	cotyledon	CaMV 35 S	synthetic *cry1EC* gene	*hph*	leaf feeding bioassay on *Spodoptera litura*	82
JL-24	*A. tumefaciens* strain C58	cotyledon	CaMV 35 S *A. thaliana* oleosin	rice chitinase maize phytoene synthase	*none* *none*	none none	71
Golden and BARI-2000	*A. tumefaciens* strain LB4404	cotyledonary node	CaMV 35 S	rice chitinase-3	*hph*	inoculation with *cerospora arachidicola*	80

Golden and BARI-2000	*A. tumefaciens* strain LBA4404	cotyledonary node	CaMV 35 S	*AtNHXI*	*npt*II	green house salt and drought tolerance	74
N/A	*A. tumefaciens* strain EHA105	embryo leaflets	Peanut oleosin	Ure B	*npt*II	none	83
New Mexico Valencia A	*A. tumefaciens* strain EHA104	cotyledon	drought inducible SARK	Isopentenyltransferase	*npt*II	Field drought tolerance test	75
Georgia Green	*A. tumefaciens* strain EHA105	hypocotyl	CaMV 35 S	*arah* 2 RNAi	*npt*II	Allergenicity by human Ig E	63
Okrun	Biolistic	somatic embyro	CaMV 35 S	rice chitinase Alfalfa glucanase	*hph*	fungal resistance and agronomic traits evaluated in a 3-year field study	79
Georgia Green	Biolistic	somatic embyro	CaMV 35 S	*arah* 2 RNAi	*hph*	Allergenicity by human Ig E	85
Georgia Green	Biolistic	somatic embyro	CaMV 35 S	chloroperoxidase	*hph*	In vitro and in situ *A. flavus* inoculation	81
JL-24	Biolistic	somatic embyro	CaMV 35 S	Bluetongue VP2	*npt*II	none	84
Georgia Green	Biolistic	somatic embyro	CaMV 35 S	Bcl-xL	*hph*	In vitro paraquat assay	76
Wilson, Perry, NC-7	Biolistic	somatic embyro	CaMV 35 S	Barley oxlate oxidase	*hph*	Field evaluation of transgenic lines for Sclerotinia minor resistance	77-78

Besides these applications in combating environmental challenges, peanut transformation has applications in vaccine development and peanut allergen silencing. Immunogen Ure B against the human bacterial pathogen *Helicobacter pylori* was overexpressed in peanut that potentially could be used as an oral vaccine [83]. The VP2 gene coding for the outer capsid of bluetongue virus (a sheep pathogen) was transformed in peanut [84]. In both cases, transgenic peanut has not been tested for the effectiveness of vaccination. Peanut endogenous proteins Ara h 2 and Ara h 6 were shown to be potent peanut allergens. Both were silenced by introducing an RNAi construct targeting homologous coding sequence, and human IgE binding to these proteins was greatly reduced in transgenic lines [63,85]. Ara h 2 was shown to have some trypsin inhibitor function [86], but silencing Ara h 2 did not promote *Aspergillus flavus* fungal growth. Collateral changes of proteins such as elevation in Ara h 10 (oleosin), 13-lipoxygenase and Ahy-3 (arachin) and decrease in conarachin among Ara h 2 silenced lines were identified by quantitative proteomics [87]. However, to date no released peanut cultivars are transgenic. There is public resistance to GMO food crops, particularly in the European countries, and it is very costly to meet the regulatory requirements for the release of GMO cultivars. In addition, there are significant issues regarding freedom to operate related to patented technologies.

Tilling

New sources of genetic variation can also be generated by TILLING (targeting induced local lesions in genomes) [88]. Tilling is a reverse genetic technique that requires knowledge of gene sequences since mutants are detected by screening for DNA sequence changes rather than phenotypic differences (forward genetics). A peanut TILLING population of over 3400 mutant lines from the cultivar Tifrunner [89] was generated using chemical (ethylmethane sulfonate -EMS) mutagenesis and screened for mutations in six genes [90]. This resulted in the discovery of gene knockouts or functional mutations in genes encoding the major allergen proteins, Ara h 1 and Ara h 2, and one of the genes that controls the oleic to linoleic acid ratio in peanut seed (FAD2). As more sequence data become available for peanut, this TILLING population should be useful for functional genomic studies as well as to discover mutations of potential value for cultivar development.

Synthetic Allotetraploids

Peanut cultivar development would greatly benefit from simplified access to the genetic diversity available in related diploid species of *Arachis*. Of great interest are the extremely high levels of resistance to many pests and diseases that occur in wild *Arachis* [91]. Introgression of traits using conventional breeding is a long and arduous task due to the cross incompatibilities and ploidy barriers between diploid wild and tetraploid cultivated along with poor agronomic performance of interspecific material. Two pathways for introgression have been tested in peanut, one involving a triploid intermediate from hybridization of cultivated tetraploid with wild diploid and the other a synthetic tetraploid hybrid crossed with cultivated tetraploid [48,92]. The latter is the most direct route for introgressing useful genes into *A. hypogaea*. Since much higher levels of molecular polymorphism occur in diploid *Arachis* in comparison to *A. hypogaea* [3,7], the use of molecular genetic technology on hybrids between wild and cultivated lines should allow for much more rapid and efficient introgression of desirable traits while maintaining acceptable agronomic performance. A synthetic tetraploid has been derived through crosses between the two putative progenitors of *A. hypogaea* (*A. duranensis* and *A. ipaensis*) and is being used as a springboard to access the diploid gene pool to mine for disease resistance, drought tolerance, and other traits [92,93].

EXAMPLES OF MOLECULAR BREEDING IN PEANUT CULTIVAR DEVELOPMENT

The first successful example of marker assisted selection (MAS) was the introgression of nematode resistance through an amphidiploid pathway into cultivated peanut [48], and the subsequent development of a nematode resistant cultivar, NemaTAM [94]. Although this cultivar has near immunity to the peanut root-knot nematode, it is not suitable for cultivation in the Southeastern U.S. due to extreme susceptibility to *tomato spotted wilt tospovirus* (TSWV) [95]. A goal of our peanut breeding program was to develop a cultivar with resistance to both TSWV and the peanut root-knot nematode. We chose to pursue this objective using phenotypic selection because the early markers were expensive, low throughput, and produced a significant amount of inaccurate data. Using a conventional breeding approach we produced 'Tifguard', the first peanut cultivar with high levels of resistance to both the peanut root-knot nematode and TSWV [95].

The next goal of the breeding program was to combine the high oleic fatty acid trait with resistance to both the peanut root-knot nematode and TSWV. Due to recent advances in molecular marker technology we decided to pursue this goal using a backcross breeding program accelerated by MAS. Research by Chu *et al.* [39] and Nagy *et al.* [40] resulted in the development of molecular markers for nematode resistance that can be used in high throughput systems and are more amenable for large breeding populations. Chu *et al.* [44,45] developed molecular markers for both genes which control the high oleic fatty acid trait in peanut. To achieve our breeding goal, Tifguard was used as the recurrent female parent and two high oleic cultivars were used as donor parents for the high O/L trait. 'Tifguard High O/L' was generated through three rounds of backcrossing using as the pollen donors BC_nF_1 progenies selected with molecular markers for these two traits. The high O/L trait is recessive but the use of a co-dominant molecular marker allowed backcrossing with heterozygous lines. Selfed BC_3F_2 plants yielded marker-homozygous individuals identified as Tifguard High O/L. Use of this MAS backcross breeding procedure compressed the hybridization and selection phases of the cultivar development process to less than 3 years [50], allowing for more rapid initiation of preliminary yield trials.

It is anticipated that more peanut cultivars developed using molecular technology will be released in the near future. Recent research has resulted in the development of several molecular markers or QTLs that should be useful for peanut cultivar development. Stalker *et al.* [96] presented a table with 14 references documenting molecular markers associated with traits in peanut, and efforts are ongoing to use the molecular markers associated with the QTL for leaf rust to incorporate leaf rust resistance into three elite cultivars at ICRISAT, India [23].

CONCLUDING COMMENTS

Efforts to shepherd initiatives for increased research on peanut genomics at the 2001 U.S. Legume Crops Genomics Workshop and at subsequent meetings of the International Peanut Genome Consortium have been described by Stalker *et al.* [96] and Feng *et al.* [20]. Recent updates are posted on http://www.peanutbioscience.com. These efforts have resulted in quantum leaps of knowledge about the peanut genome, and have facilitated ongoing marker assisted breeding programs. These efforts have also stimulated the development of molecular genetic tools and RIL populations that should result in additional quantum leaps of knowledge. In addition, these efforts have laid the foundation for plans to sequence the peanut genome in the near future. This should result in the development of additional molecular tools that will greatly advance peanut cultivar development.

REFERENCES

1. Jackson, S.A.; Iwata, A.; Lee, S.H.; Schmutz, J.; Shoemaker, R. Sequencing crop genomes: Approaches and applications. *New Phytol.* 2011, *191*, 915–925.
2. Guo, B.; Chen, C.; Chu, Y.; Holbrook, C.C.; Ozias-Akins, P.; Stalker, H.T. Advances in genetics and genomics for sustainable peanut production. In *Sustainable Agriculture and New Biotechnologies*; Benkeblia, N., Ed.; CRC Press: Boca Raton, FL, USA, 2011a; pp. 341–367.
3. Halward, T.M.; Stalker, H.T.; Larue, E.A.; Kochert, G. Genetic variation detectable with molecular markers among unadapted germ-plasm resources of cultivated peanut and related wild species. *Genome* 1991, *34*, 1013–1020.
4. Halward, T.; Stalker, T.; Larue, E.; Kochert, G. Use of single-primer DNA amplifications in genetic studies of peanut (*Arachis hypogaea* L.). *Plant Mol. Biol.* 1992, *18*, 315–325.
5. He, G.; Prakash, C.S. Identification of polymorphic DNA markers in cultivated peanut (*Arachis hypogaea* L.). *Euphytica* 1997, *97*, 143–149.

6. Herselman, L. Genetic variation among southern african cultivated peanut (*Arachis hypogaea* L.) genotypes as revealed by AFLP analysis. *Euphytica* 2003, *133*, 319–327.
7. Kochert, G.; Halward, T.; Branch, W.D.; Simpson, C.E. RFLP variability in peanut (*Arachis hypogaea* L.) cultivars and wild species. *Theor. Appl. Genet.* 1991, *81*, 565–570.
8. Gimenes, M.A.; Lopes, C.R.; Galgaro, M.L; Valls, J.F.M.; Kochert, G. RFLP analysis of genetic variation in species of section *Arachis*, genus *Arachis* (Leguminosae). *Euphytica* 2002, *123*, 421–429.
9. Milla-Lewis, S.R.; Isleib, T.G.; Stalker, H.T.; Scoles, G.J. Taxonomic relationships among *Arachis* sect. *Arachis* species as revealed by AFLP markers. *Genome* 2005, *48*, 1–11.
10. Paik-Ro, O.G.; Smith, R.L.; Knauft, D.A. Restriction fragment length polymorphism evaluation of six peanut species within the *Arachis* section. *Theor. Appl. Genet.* 1992, *84*, 201–208.
11. Stalker, H.T.; Phillips, T.D.; Murphy, J.P.; Jones, T.M. Variation of isozyme patterns among *Arachis* species. *Theor. Appl. Genet.* 1994, *87*, 746–755.
12. Dwivedi, S.L.; Crouch, J.H.; Nigam, S.N.; Ferguson, M.E.; Paterson, A.H. Molecular breeding of groundnut for enhanced productivity and food security in the semi-arid tropics: Opportunities and challenges. *Adv. Agron.* 2003, *80*, 153–221.
13. Holbrook, C.C.; Stalker, H.T. Peanut breeding and genetic resources. In *Plant Breeding Reviews*; Janick, J., Ed.; John Wiley & Sons, Inc.: Hoboken, NJ, USA, 2003; Volume 22, pp. 297–356.
14. Stalker, H.T.; Mozingo, L.G. Molecular markers of *Arachis* and marker-assisted selection. *Peanut Sci.* 2001, *28*, 117–123.
15. Varshney, R.K.; Bertioli, D.J.; Moretzsohn, M.C.; Vedez, V.; Krishnamurthy, L.; Aruna, R.; Nigam, S.N.; Moss, B.J.; Seetha, K.; Ravi, K.; *et al.* The first SSR-based genetic linkage map for cultivated groundnut (*Arachis hypogaea* L.). *Theor. Appl. Genet.* 2009, *118*, 729–739.
16. Hong, B.; Chen, X.P.; Liang, X.O.; Liu, H.Y.; Zhou, G.Y.; Li, S.X.; Wen, S.J.; Holbrook, C.C.; Guo, B.Z. A SSR-based composite genetic linkage map for cultivated peanut (*Arachis hypogaea* L.) genome. *BMC Plant Biol.* 2010, *10*, 17.

17. Qin, H.; Feng, S.; Chen, C.; Guo, Y.; Knapp, S.; Culbreath, A.; He, G.; Wang, M.L.; Zhang, X.; Holbrook, C.C.; Ozias-Akins, P.; Guo, B. An integrated genetic linkage map of cultivated peanut (*Arachis hypogaea* L.) constructed from two RIL populations. *TAG* 2012. (in press).
18. Sujay, V.; Gowda, M.V.C.; Pandey, M.K.; Bhat, R.S.; Khedikar, Y.P.; Nadaf, H.L.; Gautami, B.; Sarvamangala, C.; Lingaraju, S.; Radhakrishnan, T.; *et al.* Quantitative trait locus analysis and construction of consensus genetic map for foliar disease resistance based on two recombinant inbred line populations in cultivated groundnut (*Arachis hypogaea* L.). *Mol. Breed.* 2011. (in press).
19. Gautami, B.; Pandey, M.K.; Vadez, V.; Nigam, S.N.; Ratnakumar, P.; Krishnamurthy, L.; Radhakrishnan, T.; Gowda, M.V.C.; Narasu, M.L.; Hoisington, D.A.; *et al.* Quantitative trait locus analysis and construction of consensus genetic map for drought tolerance traits based on three recombinant inbred line populations in cultivated groundnut (*Arachis hypogaea* L.). *Mol. Breed.* 2011. (in press).
20. Feng, S.; Wang, X.; Dang, P.M.; Zhang, X.; Holbrook, C.C.; Culbreath, A.K.; Wu, Y.; Guo, B. Peanut (*Arachis hypogaea*) expressed sequence tag (EST) project: Progress and application. *Comp. Funct. Genomics* 2012. (in press).
21. Payton, P.; Kottapalli, K.R.; Rowland, D.; Faircloth, W.; Guo, B.Z.; Burow, M.; Puppala, N.; Gallo, M. Gene expression profiling in peanut using high density oligonucleotide microarrays. *BMC Genom.* 2009, *10*, 265.
22. Guo, B.Z.; Fedorova, N.D.; Chen, X.; Wan, C.H.; Wang, W.; Nierman, W.C.; Bhatnagar, D.; Yu, J. Gene expression profiling and identification of resistance genes to *Aspergillus flavus* infection in peanut through EST and microarray strategies. *Toxins* 2011b, *3*, 737–753.
23. Pandey, M.K.; Monyo, E.; Ozias-Akins, P.; Liang, X.; Guimarães, P.; Nigam, S.N.; Upadhyaya, H.D.; Janila, P.; Zhang, X.; Guo, B.; *et al.* Advances in *Arachis* genomics for peanut improvement. *Biotechnol. Adv.* 2011a. (in press).
24. Moretzsohn, M.C.; Hopkins, M.S.; Mitchell, S.E.; Kresovich, S.; Valls, J.F.; Ferreira, M.E. Genetic diversity of peanut (*Arachis hypogaea* L.) and its wild relatives based on the analysis of

hypervariable regions of the genome. *BMC Plant Biol.* 2004, *4*, 11.

25. Liang, X.Q.; Chen, X.P.; Hong, Y.B.; Liu, H.Y.; Zhou, G.Y.; Li, S.X.; Guo, B.Z. Utility of EST-derived SSR in cultivated peanut (*Arachis hypogaea* L.) and *Arachis* wild species. *BMC Plant Biol.* 2009, *9*, 35.
26. Yuan, M.; Gong, L.M.; Meng, R.H.; Li, S.L.; Dang, P.; Guo, B.Z.; He, G.H. Development of trinucleotide (GGC)n SSR markers in peanut (*Arachis hypogaea* l.). *Electron. J. Biotechnol.* 2010, *13*, 6.
27. Pandey, M.; Gautami, B.; Jayakumar, T.; Sriswathi, M.; Upahyaya, H.D.; Gowda, M.V.C.; Radhakrishnan, T.; Bertioli, D.J.; Knapp, S.J.; Cook, D.R.; *et al.* Highly informative genic and genomic SSR markers to facilitate molecular breeding in cultivated groundnut (*Arachis hypogaea*). *Plant Breed.* 2011b. (in press).
28. Hopkins, M.S.; Casa, A.M.; Wang, T.; Mitchell, S.E.; Dean, R.E.; Kochert, G.D.; Kresovich, S. Discovery and characterization of polymorphic simple sequence repeats (SSRs) in peanut. *Crop Sci.* 1999, *39*, 1243–1247.
29. He, G.; Meng, R.; Newman, M.; Gao, G.; Pittman, R.N.; Prakash, C.S. Microsatellites as DNA markers in cultivated peanut (*Arachis hypogaea* L.). *BMC Plant Biol.* 2003, *3*, 3.
30. Ferguson, M.E.; Burow, M.D.; Schulze, S.R.; Bramel, P.J.; Paterson, A.H.; Kresovich, S.; Mitchell, S. Microsatellite identification and characterization in peanut (A. *hypogaea* l.). *Theor. Appl. Genet.* 2004, *108*, 1064–1070.
31. Wang, C.T.; Yang, X.D.; Chen, D.X.; Yu, S.L.; Liu, G.Z.; Tang, Y.Y.; Xu, J.Z. Isolation of simple sequence repeats from groundnut. *Electron. J. Biotechnol.* 2007, *10*, 473–480.
32. Cuc, L.M.; Mace, E.S.; Crouch, J.H.; Quang, V.D.; Long, T.D.; Varshney, R.K. Isolation and characterization of novel microsatellite markers and their application for diversity assessment in cultivated groundnut (*Arachis hypogaea*). *BMC Plant Biol.* 2008, *8*, 55.
33. Kochert, G.; Stalker, H.T.; Gimenes, M.; Galgaro, L.; Lopes, C.R.; Moore, K. RFLP and Cytogenetic evidence on the origin and evolution of allotetraploid domesticated peanut, *Arachis hypogaea* (Leguminosae). *Am. J. Bot.* 1996, *83*, 1282–1291.

34. Mace, E.S.; Phong, D.T.; Upadhyaya, H.D.; Chandra, S.; Crouch, J.H. SSR analysis of cultivated groundnut (*Arachis hypogaea* L.) germplasm resistant to rust and late leaf spot diseases. *Euphytica* 2006, *152*, 317–330.
35. Mondal, S.; Badigannavar, A.M. Molecular diversity and association of SSR markers to rust and late leaf spot resistance in cultivated groundnut (*Arachis hypogaea* L.). *Plant Breed.* 2010, *129*, 68–71.
36. Khedikar, Y.P.; Gowda, M.V.C.; Sarvamangala, C.; Patgar, K.V.; Upadhyaya, H.D.; Varshney, R.K. A QTL study on late leaf spot and rust revealed one major QTL for molecular breeding for rust resistance in groundnut (*Arachis hypogaea* L.). *Theor. Appl. Genet.* 2010, *121*, 971–984.
37. Burow, M.D.; Simpson, C.E.; Paterson, A.H.; Starr, J.L. Identification of peanut (*Arachis hypogaea* l.) RAPD markers diagnostic of root-knot nematode (*Meloidogyne arenaria* (Neal) Chitwood) resistance. *Mol. Breed.* 1996, *2*, 369–379.
38. Garcia, G.M.; Stalker, H.T.; Shroeder, E.; Kochert, G. Identification of RAPD, SCAR, and RFLP markers tightly linked to nematode resistance genes introgressed from *Arachis cardenasii* into *Arachis hypogaea*. *Genome* 1996, *39*, 836–845.
39. Chu, Y.; Holbrook, C.C.; Timper, P.; Ozias-Akins, P. Development of a PCR-based molecular marker to select for nematode resistance in peanut. *Crop Sci.* 2007a, *7*, 841–847.
40. Nagy, E.D.; Chu, Y.; Guo, Y.; Khanal, S.; Tang, S.; Li, Y.; Dong, W.B.; Timper, P.; Taylor, C.; Ozias-Akins, P.; *et al.* Recombination is suppressed in an alien introgression in peanut harboring *Rma*, a dominant root-knot nematode resistance gene. *Mol. Breed.* 2010, *26*, 357–370.
41. Herselman, L.; Thwaites, R.; Kimmins, F.M.; Courtois, B.; Van Der Merwe, P.J.A.; Seal, S.E. Identification and mapping of AFLP markers linked to peanut (*Arachis hypogaea* L.) resistance to the aphid vector of groundnut rosette disease. *Theor. Appl. Genet.* 2004, *109*, 1426–1433.
42. Ravi, K.; Vadez, V.; Isobe, S.; Mir, R.R.; Guo, Y.; Nigam, S.N.; Gowda, M.V.C.; Radhakrishnan, T.; Bertioli, D.J.; Knapp, S.J.; *et al.* Identification of several small main-effect QTLs and a large number of epistatic QTLs for drought tolerance related traits in

groundnut (*Arachis hypogaea* L.). *Theor. Appl. Gen.* 2011, *122,* 1119–1132.

43. Selvaraj, M.G.; Narayana, M.; Schubert, A.M.; Ayers, J.L.; Baring, M.R.; Burow, M.D. Identification of QTLs for pod and kernel traits in cultivated peanut by bulked segregant analysis. *Electron. J. Biotechnol.* 2009, *12,* 13.
44. Chu, Y.; Ramos, L.; Holbrook, C.C.; Ozias-Akins, P. Frequency of a loss-of-function mutation inoleoyl-PC desaturase (ahFAD2A) in the mini-core of the U.S. peanut germplasm collection. *Crop Sci.* 2007b, *47,* 2372–2378.
45. Chu, Y.; Holbrook, C.C.; Ozias-Akins, P. Two alleles of ahFAD2B control the high oleic acid trait in cultivated peanut. *Crop Sci.* 2009, *49,* 2029–2036.
46. Sarvamangala, C.; Gowda, M.V.C.; Varshney, R.K. Identification of quantitative trait loci for protein content, oil content and oil quality for groundnut (*Arachis hypogaea* L.). *Field Crops Res.* 2011, *122,* 49–59.
47. Gowda, M.V.C.; Motagi, B.N.; Naidu, G.K.; Diddimani, S.B.; Sheshagiri, R. Gpbd 4: A spanish bunch groundnut genotype resistant to rust and late leaf spot. *Int. Arachis Newslett.* 2002, *22,* 29–32.
48. Simpson, C.E. Use of wild arachis species/introgression of genes into A. hypogaea L. *Peanut Sci.* 2001, *28,* 114–116.
49. Varshney, R.K.; Dubey, A. Novel genomic tools and modern genetic and breeding approaches for crop improvement. *J. Plant Biochem. Biotechnol.* 2009, *18,* 127–138.
50. Chu, Y.; Wu, C.L.; Holbrook, C.C.; Tillman, B.; Person, G.; Ozias-Akins, P. Marker-assisted selection to pyramid nematode resistance and the high oleic trait in peanut. *Plant Genome* 2011, *4,* 110–117.
51. Varshney, R.K.; Close, T.J.; Singh, N.K.; Hoisington, D.A.; Cook, D.R. Orphan legume crops enter the genomics era! *Curr. Opin. Plant Biol.* 2009b, *12,* 202–210.
52. Varshney, R.K.; Glaszmann, J.C.; Leung, H.; Ribaut, J.M. More genomic resources for less-studied crops. *Trends Biotechnol.* 2010, *28,* 452–460.

53. Ozias-Akins, P.; Schnall, J.A.; Anderson, W.F.; Singsit, C.; Clemente, T.E.; Adang, M.J.; Weissinger, A.K. Regeneration of transgenic peanut plants from stably transformed embryogenic callus. *Plant Sci.* 1993, *93*, 185–194.

54. Klein, T.M.; Wolf, E.D.; Wu, R.; Sanford, J.C. High-velocity microprojectiles for delivering nucleic acids into living cells. *Nature* 1987, *327*, 70–73.

55. Singsit, C.; Adang, M.J.; Lynch, R.E.; Anderson, W.F.; Wang, A.M.; Cardineau, G.; Ozias-Akins, P. Expression of a *Bacillus thuringiensis cryIA(c)* gene in transgenic peanut plants and its efficacy against lesser cornstalk borer. *Transgen. Res.* 1997, *6*, 169–176.

56. Palmer, A.G.; Gao, R.; Maresh, J.; Erbil, W.K.; Lynn, D.G. Chemical biology of multi-host/pathogen interactions: Chemical perception and metabolic complementation. *Annu. Rev. Phytopathol.* 2004, *42*, 439–464.

57. Pitzschke, A.; Hirt, H. New insights into an old story: Agrobacterium-induced tumour formation in plants by plant transformation. *EMBO J.* 2010, *29*, 1021–1032.

58. Li, Z.; Jarret, R.L.; Demski, J.W. Engineered resistance to tomato spotted wilt virus in transgenic peanut expressing the viral nucleocapsid gene. *Transgen. Res.* 1997, *6*, 297–305.

59. Sharma, K.K.; Anjaiah, V. An efficient method for the production of transgenic plants of peanut (*Arachis hypogaea* L.) through *Agrobacterium tumefaciens*-mediated genetic transformation. *Plant Sci.* 2000, *159*, 7–19.

60. Swathi Anuradha, T.; Divya, K.; Jami, S.K.; Kirti, P.B. Transgenic tobacco and peanut plants expressing a mustard defensin show resistance to fungal pathogens. *Plant Cell Rep.* 2008, *27*, 1777–1786.

61. Ozias-Akins, P. Peanut. In *Biotechnology in Agriculture and Forestry*; Pua, E.C., Davey, M.R., Eds.; Springer-Verlag: Berlin/Heidlberg, Germany, 2005; pp. 81–105.

62. Ozias-Akins, P.; Gill, R. Progress in the development of tissue culture and transformation methods applicable to the production of transgenic peanut. *Peanut Sci.* 2001, *28*, 123–131.

63. Dodo, H.W.; Konan, K.N.; Chen, F.C.; Egnin, M.; Viquez, O.M. Alleviating peanut allergy using genetic engineering: The silencing of the immunodominant allergen Ara h 2 leads to its significant reduction and a decrease in peanut allergenicity. *Plant Biotechnol. J.* 2008, *6*, 135–145.
64. Tie, W.; Zhou, F.; Wang, L.; Xie, W.; Chen, H.; Li, X.; Lin, Y. Reasons for lower transformation efficiency in indica rice using agrobacterium tumefaciens-mediated transformation: Lessons from transformation assays and genome-wide expression profiling. *Plant Mol. Biol.* 2011, *78*, 1–18.
65. Branch, W.D. Registration of 'Georgia Green' peanut. *Crop Sci.* 1996, *36*, 806.
66. Gorbet, D.W.; Tillman, B.L. Registration of 'Florida-07' peanut. *J. Plant Reg.* 2009, *3*, 138–142.
67. Burns, S.P.; Gallo, M.; Tillman, B.L. Expansion of a direct shoot organogenesis system in peanut (*Arachis hypogaea* L.) to include US cultivars. *In Vitro Cell. Dev. Biol. Plant* 2011. (in press).
68. Swathi Anuradha, T.; Jami, S.K.; Datla, R.S.; Kirti, P.B. Genetic transformation of peanut (*Arachis hypogaea* L.) using cotyledonary node as explant and a promoterless gus::*npt*II fusion gene based vector. *J. Biosci.* 2006, *31*, 235–246.
69. Chenault, K.D.; Gallo, M.; Ozias-Akins, P.; Srivastava, P. Peanut. In *Compendium of Transgenic Crop Plants: Transgenic Oilseed Crop*; Wiley, J. & Sons: Oxford, UK, 2008; pp. 169–198.
70. Ozias-Akins, P. Peanut. In *Biotechnology in Agriculture and Forestry 61—Transgenic Crops VI*; Pua, E.C., Davey, M.R., Eds.; Springer: Berlin, Germany, 2007; pp. 81–105.
71. Bhatnagar, M.; Prasad, K.; Bhatnagar-Mathur, P.; Narasu, M.L.; Waliyar, F.; Sharma, K.K. An efficient method for the production of marker-free transgenic plants of peanut (*Arachis hypogaea* L.). *Plant Cell Rep.* 2010, *29*, 495–502.
72. Bhatnagar-Mathur, P.; Devi, M.J.; Reddy, D.S.; Lavanya, M.; Vadez, V.; Serraj, R.; Yamaguchi-Shinozaki, K.; Sharma, K.K. Stress-inducible expression of at DREB1A in transgenic peanut (*Arachis hypogaea* L.) increases transpiration efficiency under water-limiting conditions. *Plant Cell Rep.* 2007, *26*, 2071–2082.

73. Bhatnagar-Mathur, P.; Devi, M.J.; Vadez, V.; Sharma, K.K. Differential antioxidative responses in transgenic peanut bear no relationship to their superior transpiration efficiency under drought stress. *J. Plant Physiol.* 2009, *166*, 1207–1217.

74. Asif, M.A.; Zafar, Y.; Iqbal, J.; Iqbal, M.M.; Rashid, U.; Ali, G.M.; Arif, A.; Nazir, F. Enhanced expression of AtNHX1, in transgenic groundnut (*Arachis hypogaea* L.) improves salt and drought tolerance. *Mol. Biotechnol.* 2011, *49*, 250–256.

75. Qin, H.; Gu, Q.; Zhang, J.; Sun, L.; Kuppu, S.; Zhang, Y.; Burow, M.; Payton, P.; Blumwald, E.; Zhang, H. Regulated expression of an isopentenyltransferase gene (*IPT*) in peanut significantly improves drought tolerance and increases yield under field conditions. *Plant Cell Physiol.* 2011, *52*, 1904–1914.

76. Chu, Y.; Deng, X.Y.; Faustinelli, P.; Ozias-Akins, P. Bcl-xL transformed peanut (*Arachis hypogaea* L.) exhibits paraquat tolerance. *Plant Cell Rep.* 2008a, *27*, 85–92.

77. Livingstone, D.M.; Hampton, J.L.; Phipps, P.M.; Grabau, E.A. Enhancing resistance to sclerotinia minor in peanut by expressing a barley oxalate oxidase gene. *Plant Physiol.* 2005, *137*, 1354–1362.

78. Partridge-Telenko, D.E.; Hu, J.; Livingstone, D.M.; Shew, B.B.; Phipps, P.M.; Grabau, E.A. Sclerotinia blight resistance in virginia-type peanut transformed with a barley oxalate oxidase gene. *Phytopathology* 2011, *101*, 786–793.

79. Chenault, K.D.; Melouk, H.A.; Payton, M.E. Effect of anti-fungal transgene(s) on agronomic traits of transgenic peanut lines grown under field conditions. *Peanut Sci.* 2006, *33*, 12–19.

80. Iqbal, M.M.; Nazir, F.; Ali, S.; Asif, M.A.; Zafar, Y.; Iqbal, J.; Ali, G.M. Over expression of rice chitinase gene in transgenic peanut (*Arachis hypogaea* L.) improves resistance against leaf spot. *Mol. Biotechnol.* 2011. (in press).

81. Niu, C.; Akasaka-Kennedy, Y.; Faustinelli, P.; Joshi, M.; Rajasekaran, K.; Yang, H.; Chu, Y.; Cary, J.; Ozias-Akins, P. Antifungal activity in transgenic peanut (*Arachis hypogaea L.*) conferred by a nonheme chloroperoxidase gene. *Peanut Sci.* 2009, *36*, 126–132.

82. Tiwari, S.; Mishra, D.K.; Singh, A.; Singh, P.K.; Tuli, R. Expression of a synthetic *cry1EC* gene for resistance against *Spodoptera*

litura in transgenic peanut (*Arachis hypogaea* L.). *Plant Cell Rep.* 2008, *27*, 1017–1025.

83. Yang, C.Y.; Chen, S.Y.; Duan, G.C. Transgenic peanut (*Arachis hypogaea* L.) expressing the urease subunit B gene of *Helicobacter pylori. Curr. Microbiol.* 2011, *63*, 387–391.
84. Athmaram, T.N.; Bali, G.; Devaiah, K.M. Integration and expression of bluetongue VP2 gene in somatic embryos of peanut through particle bombardment method. *Vaccine* 2006, *24*, 2994–3000.
85. Chu, Y.; Faustinelli, P.; Ramos, M.L.; Hajduch, M.; Thelen, J.J.; Maleki, S.J.; Ozias-Akins, P. Reduction of IgE binding and non-promotion of *Aspergillus flavus* fungal growth by simultaneously silencing Ara h 2 and Ara h 6 in peanut. *J. Agric. Food Chem.* 2008b. in press.
86. Maleki, S.J.; Viquez, O.; Jacks, T.; Dodo, H.; Champagne, E.T.; Chung, S.Y.; Landry, S.J. The major peanut allergen, Ara h 2, functions as a trypsin inhibitor, and roasting enhances this function. *J. Allergy Clin. Immunol.* 2003, *112*, 190–195.
87. Stevenson, S.E.; Chu, Y.; Ozias-Akins, P.; Thelen, J.J. Validation of gel-free, label-free quantitative proteomics approaches: Applications for seed allergen profiling. *J. Proteomics* 2009, *72*, 555–566.
88. McCallum, C.; Comai, L.; Greene, E.; Henidoff, S. Targeting induced local lesions in genomes (Tilling) for plant functional genomics. *Plant Physiol.* 2000, *123*, 439–442.
89. Holbrook, C.C.; Culbreath, A.K. Registration of 'Tifrunner' peanut. *J. Plant Reg.* 2007, *1*, 124.
90. Knoll, J.E.; Ramos, M.L.; Zeng, Y.; Holbrook, C.C.; Chow, M.; Chen, S.; Maleki, S.; Bhattacharya, A.; Ozias-Akins, P. TILLING for allergen reduction and improvement of quality traits in peanut (*Arachis hypogaea* L.). *BMC Plant Biol.* 2011, *11*, 81.
91. Stalker, H.T.; Simpson, C.E. Germplasm resources in *Arachis*. In *Advances in Peanut Science*; Pattee, H.E., Stalker, H.T., Eds.; Amer. Peanut Res. and Educ. Soc.: Stillwater, OK, USA, 1995; pp. 14–53.

92. Bertioli, D.J.; Seijo, G.; Freitas, F.O.; Valls, J.F.M.; Leal-Bertioli, S.C.M.; Moretzsohn, M.C. An overview of peanut and its wild relatives. *Plant Gen. Res.-Char. Util.* 2011, *9*, 134–149.

93. Foncéka, D.; Hodo-Abalo, T.; Rivallan, R.; Faye, I.; Sall, M.N.; Ndoye, O.; Fávero, A.P.; Bertioli, D.J.; Glaszmann, J.C.; Coutois, B.; *et al.* Genetic mapping of wild introgressions into cultivated peanut: A way toward enlarging the genetic basis of a recent allotetraploid. *BMC Plant Biol.* 2009, *9*, 103.

94. Simpson, C.E.; Burrow, M.D.; Patterson, A.H.; Starr, J.L.; Church, G.T. Registration of 'NemaTAM' peanut. *Crop Sci.* 2003, *43*, 1561.

95. Holbrook, C.C.; Timper, P.; Culbreath, A.K.; Kvien, C.K. Registration of 'Tifguard' peanut. *J. Plant Reg.* 2008, *2*, 92–94.

96. Stalker, H.T.; Weissinger, A.K.; Milla-Lewis, S.; Holbrook, C.C. Genomics: An evolving science in peanut. *Peanut Sci.* 2009, *36*, 2–10.

Chapter 4

Mutagenesis: A Useful Tool for the Genetic Improvement of the Cultivated Peanut (Arachis hypogaea L.)

Chuan Tang Wang[1], Yue Yi Tang[1], Xiu Zhen Wang[1], Qi Wu[1], Hua Yuan Gao[2], Tong Feng[2], Jun Wei Su[3], Shu Tao Yu[3], Xian Lan Fang[4], Wan Li Ni[5], Yan Sheng Jiang[6], Lang Qian[7] and Dong Qing Hu[8]

[1]Shandong Peanut Research Institute (SPRI), Qingdao,, China

[2]Institute of Economic Plants, Jilin Academy of Agricultural Sciences, Gongzhuling,, China

[3]Sandy Land Amelioration and Utilization Research Institute of Liaoning,Fuxin,, China

[4]South Jiangxi Academy of Sciences, Ganzhou,, China

[5] Institute of Crop Plants, Anhui Academy of Agricultural Sciences, Hefei,, China

[6]Weifang Academy of Agricultural Sciences, Weifang,, China

[7]Dalian Academy of Agricultural Sciences, Dalian,, China

[8]Qingdao Entry-Exit Inspection and Quarantine Bureau, Qingdao,, China

INTRODUCTION

Ranking 4th among world oilseed crops and 13th among food crops, the cultivated peanut (*Arachis hypogaea* L.) is an important cash crop currently grown in over eighty countries/regions from 40°N to 40°S across tropical and warm temperate regions [1,2]. Its seeds contain about 50% oil and 25% protein, and the crop is thus deemed as a rich source of edible oil and dietary protein. In developing countries, a large portion of peanuts are crushed for edible oil [3]. Food uses of peanut are predominant in developed nations, where high oleate, high protein and reduced fat peanuts are most preferred, as high oleate not only means better keeping quality, but also brings about multiple health benefits, for example, reduced risk of cardiovascular disease, increased sensitivity to insulin, preventative effects on tumorigenesis, and amelioration of some inflammatory diseases [3,4].

Breeding for high yield has been and will continue to be one of the most important objectives of any peanut genetic improvement programs. For quite a long time, intraspecific hybridization (IH) has been the major breeding method of peanut, through which a sizable number of peanut cultivars with high yields have been released. The narrow gene base of the peanut cultigen caused by reproductive isolation from wild *Arachis* relatives has rendered the genetic improvement of the cultivated peanut through IH more and more difficult. To obtain fertile interspecific hybrids, in most cases special technical measures have to be taken to cope with incompatible obstacles and/or ploidy difference in wide crosses [5,6]. Furthermore, backcrossing is needed to break undesirable linkages. Wide segregation range in peanut interspecific derivatives and linkage drags make the breeding process using wild *Arachis* species lengthy and tedious. In contrast, traits in mutagenized populations tend to stabilize easily. Recent years, in peanut breeding, much attention has therefore been paid to mutagenesis.

Developing peanut cultivars with improved quality has long been proposed as a major breeding objective of the crop. With the depletion of fossil fuel, interest in peanut as a source of renewable energy is growing. Peanut genotypes with both high oil and high oleate are considered most suitable for biodiesel production. Yet little progress had been made in peanut quality improvement, especially for oil and protein content during the past several decades, largely due to the limited genetic variability within the cultivated peanut gene pool as well as unavailability of simple, rapid and cost-effective selection techniques. Conventional analytical procedures for oil, protein and fatty acid determination are destructive, costly and time-consuming, unsuitable for handling large samples frequently encountered in the process of breeding [7]. Fortunately, the stagnant situation is being changed as the wide application of mutagenesis in peanut breeding and the development of near infrared spectroscopy calibration equations for peanut by several authors [8-14].

Using sun-dried peanut seed we have developed NIRS (near infrared reflectance spectroscopy)calibration equations predictive of main quality characters of bulk seed samples and intact single seeds, making it possible to screen a large number of peanut seeds for multiple quality traits rapidly, simultaneously and non-destructively [13,14].

This chapter summarized the recent progress in peanut mutagenesis for yield and quality made by scientists from our research group.

DEVELOPMENT OF HIGH-YIELDING PEANUT MUTANTS THROUGH CHEMICAL MUTAGENESIS

Flower Injection of Ethyl Methane Sulfonate (EMS)

Through injection of 0.3% EMS into flowers of Huayu 16 at 9:00-9:30 a.m. and subsequent selection, we were able to develop a high-yielding peanut cultivar - Huayu 40 [15,16](Table 1).

Huayu 40 has an erect growth habit and sequential branching pattern. As compared with its wild type (Huayu 16), Huayu 40 possesses faster growing and darker green foliage [16]. A study conducted at flowering and pegging stage in 2009 showed that leaf water content, chlorophyll a and b content of Huayu 40 were significantly higher than those of Huayu 16 [15]. IT-ISJ (Intron-Targeted Intron-Exon Splice Conjunction) profiling using 7 primer pairs resulted in 8 bands that could differentiate Huayu 40 and Huayu 16 [15,17]. While Huayu 16 has white and yellow inner seed coat color, Huayu 40 only has white inner seed coat color. At Shandong Peanut Research Institute (SPRI), Huayu 40 showed a yield increase of about 5% over Huayu 16 [15] (Table 1). Basal pod setting around central axes and high peg strength minimizes pod losses at harvest [15,16].

The cultivar, also known as 08-test-A2, was approved for release in Anhui province in 2011 and in Jilin province in 2012 [15,16]. In summer sowing in Anhui province, Huayu 40 matured in 116 days, and produced an average pod yield of 3970.8 kg ha^{-1} at 3 locations, outyielding the local control Luhua 8 by 14.6%, ranking first among the 11 entries in the test [16]. In Jilin provincial peanut cultivar evaluation test conducted in 2010 and 2011, Huayu 40 matured in 124 days. It produced an average pod yield of 3385 kg ha^{-1}, 22.32% over the local check Baisha 1016.

Table 1. Huayu 40: how it was bred

Year	Action
2001	EMS injection and seed harvest
2002	Cultivation and harvest. Extremely poorly performed plants were discarded
2003-2004	Single plant selection
2005	A plant line (E7-2) with more sound mature kernels (SMK) was noticed and selected
2006-2007	Multiplication of E7-2 to obtain adequate seeds for primary yield evaluation
2008-2009	Yield evaluation in Laixi, Shandong, under the name of 08-test-A2. It showed a superiority of 24.93% (kernel yield) over Luhua 11 in 2008, 5.42% and 10.71% over Huayu 16 and Fenghua 1, respectively, in 2009.

2010-present	Continual yield evaluation and seed multiplication in Laixi, Shandong, and in Gongzhuling, Jilin. National peanut primary and secondary yield evaluation test in northern China (2010 and 2011, spring sowing), National peanut pre-release final yield evaluation test in northern China (2012, spring sowing), Jilin provincial peanut cultivar evaluation test (2010 and 2011, spring sowing), Anhui yield evaluation test (2011, summer sowing)

In 2010 Anhui yield evaluation test (2010, summer sowing), Jilin provincial peanut cultivar evaluation test (2010 and 2011, spring sowing), national peanut primary and secondary yield evaluation test in northern China (2010 and 2011, spring sowing), national peanut pre-release final yield evaluation test in northern China (201 2, spring sowing), Huayu 40 produced an average kernel yield of 3057.15 kg ha^{-1}, outyielding the national control Huayu 19 by 8.22%. Having passed the 2 years' regional evaluation test in northern China, Huayu 40 is eligible to enter the national peanut pre-release final yield evaluation test to be performed in May-September, 2012.

In a trial conducted in 2011 to select suitable peanut cultivars for northeast China, Huayu 40 showed high and stable yields across all locations (Prof Hua Yuan Gao, Hong Bo Yu, Shu Li Kang, unpublished data).

Seed Treatment with Chemical Mutagens

Other lines with high productivity derived from chemical mutagenized peanut cultivars have also been bred. In primary yield evaluation test conducted in 2011 in Laixi, Shandong, three mutant lines derived from Huayu 22 (a Virginia type peanut cultivar) seeds performed well. 11-L36, a line developed through treatment of Huayu 22 peanut seeds with 0.39% sodium azide (NaN_3), outyielded the local control Fenghua 1 by 27.04% (kernel yield). 11-L39 and 11-L40, both bred through treatment of Huayu 22 peanut seeds with 0.39% diethyl sulphate (DES), had 37.60% and 22.60% more kernel yield than Fenghua 1. These promising lines are to be tested further for yield stability over years and locations.

UTILITY OF MUTAGENESIS IN PEANUT BREEDING FOR BETTER QUALITY

NIRS to Select Quality Materials from EMS Mutagenized Peanut Populations

Previously, with the help of NIRS, a peanut plant (M_2) with elevated oleate content was selected from sodium azide mutagenized Huayu 22 seeds [14]. In autumn 2011, single seeds (M_5) with over 70% oleate content was identified by NIRS and further confirmed by GC (gas chromatography).

At SPRI, the NIRS calibration equations has also been successfully used to identify individual single peanut seeds with high oleate, high oil or high protein from EMS mutagenized peanut populations. In the experiment, 2 Virginia type peanut cultivars with desirable external traits for export, viz., LF 2 and Huayu 22 (Table 2), were chosen for mutagenic treatment with a hope to develop peanut cultivars with improved quality attributes and comparable or even higher productivity. The seeds in mesh bags (1140 seeds for each genotype) were soaked in tap water for 4 hours. Just prior to EMS treatment, 0.5%, 1.0% and 1.5% EMS solutions were prepared in 0.1 M phosphate buffered saline (PBS) (pH 7.0). The pre-soaked seeds were then treated with EMS (5ml EMS solution per seed) for 2 hours with continuous agitation. After treatment, the seeds were thoroughly washed in running water for 2 hours. Untreated seeds of LF 2 and Huayu 22 were used as controls. After 30 min of air drying, the mutagenized seeds and the untreated controls were sown (one seed per hill) in twin-row seedbeds with 80 cm bed spacing, 30 cm inter-row spacing and 16.67 cm within-row inter-plant spacing under polythene mulch on 4 May 2010 at the Nianzhitou Experimental Plots, Laixi, China. Routine cultural practices were followed [18]. Since rainfall was adequate during the crop season, no irrigation was applied. Peanut was harvested on 12 September 2010.

As shown in Table 2, EMS treatment of peanut seeds significant influenced the percentage of fertile plants, and the productivity of resultant M_1 plants as well (data not shown). With the increase of EMS concentration from 0.5 % to 1.5%, the number of fertile peanut plants harvested generally decreased; so did the number of plants producing adequate seeds as bulk seed samples for NIRS scanning (Table 2).

Oleate, protein and oil contents of sun dried peanut seeds (M_2) from individual single peanut plants (M_1) were predicted by NIRS using the calibration equations for bulk seed samples [13,14], provided that seeds were enough for the rotating sampling cup of the NIRS machine (Matrix-I, Bruker Optics, Germany). Each seed sample was measured once. Only the seeds from individual single plants predicted as with >58% oleate, >55% oil or >28% protein were used for further analysis. Individual single seeds from selected single plants were then scanned with the same NIRS machine using a small cup for a single seed. Each seed sample was measured 3 times. Oleate, oil and protein contents were predicted by NIRS equations for single intact peanut seeds [13,14]. For comparison, seed samples (at least 30 seed samples from individual single plants and at least 30 intact single seeds for each genotype) of the untreated controls were analyzed with NIRS. The number of M_1 peanut plants with >60% oleate, >55% oil, or >27% protein predicted by NIRS calibration equations for bulk seed samples and the number of single intact M_2 seeds from these plants that had over 58% oleate, >55% oil for LF 2 and >58% oil for Huayu 22, or >28% protein, predicted by NIRS calibration equations for single intact seeds were listed in Table 3 and Table 4. A large number of single seeds selected had quality traits going far beyond the variation scope of the controls (Table 5). There were marked genotypic effects on quality of M_2 peanut seeds. For LF2-derived populations, 8, 118 and 11 M_2 seeds were predicted to have >60% oleate, >55% oil, and >28% protein by NIRS, respectively; of the 3 treatments, 1.0% EMS produced the largest number of quality materials, as far as oleate, oil and protein contents were concerned (Table 3). For Huayu 22, a total of 14, 70 and 23 M_2 seeds were predicted to have >60% oleate, >58% oil, and >28% protein by NIRS, respectively; a large portion of high oleate/ protein seeds (M_2) were from 0.5% EMS treatment, while 1.5% EMS was most suitable for induction of high oil mutations (Table 4). The frequency of high oil single seeds was generally higher than that of high oleate/ protein single seeds (Table 3 and Table 4).

Table 2. No. of seeds treated with EMS (NST), no. of fertile plants harvested (NFPH) and no. of plants scanned by NIRS (NPS)

EMS (%	LF2			Huayu 22		
	NST	NFPH	NPS(%)	NST	NFPH	NPS
0.5	1140	717(62.89)*	517(45.35)	1140	839(73.60)	523(45.88)
1.0	1140	740(64.91)	413(36.23)	1140	588(51.58)	488(42.81)
1.5	1140	425(37.28)	231(20.26)	1140	422(37.02)	324(28.41)

Table 3. No. of quality plants (M_1)/seeds (M_2) identified by NIRS in mutagenized LF 2 populations

EMS (%	No. of quality plants predicted by NIRS			No. of quality seeds predicted by NIRS		
	Oleate (“/>58%)	Oil (“/>55)	Protein (“/>27%)	Oleate (“/>60%)	Oil (“/>55%)	Protein (“/>28%)
0.5	6	2	1	1	33	5
1.0	7	7	13	6	70	6
1.5	4	1	1	1	15	0

Table 4. No. of quality plants (M_1)/seeds (M_2) identified by NIRS in mutagenized Huayu 22 populations

EMS (%	No. of quality plants predicted by NIRS			No. of quality seeds predicted by NIRS		
	Oleate (“/>58%)	Oil (“/>55%)	Protein (“/>27%)	Oleate (“/>60%)	Oil (“/>58%)	Protein (“/>28%)
0.5	4	2	1	11	15	19
1.0	6	4	5	2	14	2
1.5	10	4	6	1	41	2

Table 5. Variation ranges and averages of oleate, oil and protein content in bulk seeds from single plants and in single seeds of two peanut cultivars as predicted by NIRS

Untreated control		Single plant			Single seed		
Oleate (%)		Oil (%)	Protein (%)	Oleate (%)	Oil (%)	Protein (%)	
LF 2	Min.	39.80	47.76	24.10	29.26	47.41	20.94
	Max.	51.53	51.46	26.52	39.70	53.19	27.70
	Mean	50.54	49.30	24.56	38.96	50.00	21.00
Huayu 22	Min.	49.28	47.58	23.20	30.96	51.63	20.75
	Max.	56.86	52.24	26.77	52.25	56.99	26.63
	Mean	50.12	49.97	25.90	50.30	55.50	23.82

Some seeds (M_2) with high oleate, oil or protein content predicted by NIRS were sampled and sent to Food Supervising and Testing Centre (Wuhan), China to analyze their quality by standard methods. For quality analysis by conventional means, a small seed portion of selected single seeds, distal to embryo end, weighing no less than 100 mg, was cut off and sent. Oleate, oil and protein contents were determined using GC for determination of fatty acids, Soxlet oil extraction method and Kjeldhal nitrogen determination procedure, respectively. A conversion factor of 5.46 was used to convert the amount of nitrogen to amount of proteins in the samples.

High oleate trait of 5 M_2 seeds selected by NIRS were confirmed by GC analysis; oleate content in these seeds was not less than 60.0%, significantly higher than that in untreated seeds (ck) (Table 6 andTable 5). Similarly, several high oil/protein M_2 seeds were also analyzed with conventional methods, and all of them were found to contain over 60% oil or over 29% protein, much higher than untreated controls (Table 7 and Table 8).

Table 6. Oleate content (%) in selected single peanut seeds (M_2)

EMS (%)	Cultivar	Seed serial no.	NIRS	GC
0 (ck)	LF 2	LF 2-1-15	38.1	44.2
1.0	LF 2	E2-4-256-24	60.5	62.9
1.0	LF 2	E2-4-178-18	62.4	66.4

1.0	LF 2	E2-4-83-12	63.6	64.3
0(ck)	Huayu 22	HY22-2-10	34.9	47.3
0.5	Huayu 22	E1-3-173-10	63.8	67.3

Table 7. Oil content (%) in selected single peanut seeds (M_2)

EMS (%)	Cultivar	Seed serial no.	NIRS	Soxhlet extraction
0(ck)	Huayu 22	HY22-2-10	55.5	52.82
0.5	Huayu 22	E1-3-343-24	60.2	61.49
0.5	Huayu 22	E1-3-343-25	60.8	62.38

Table 8. Protein content (%) in selected single peanut seeds (M_2)

EMS (%)	Cultivar	Seed serial no.	NIRS	Kjeldahl nitrogen determination
0 (ck)	Huayu 22	HY22-2-10	20.8	19.5
0.5	Huayu 22	E1-3-180-6	29.0	31.35
0.5	Huayu 22	E1-3-180-10	28.6	30.49

M_2 seeds that were identified as with high oil, protein or oleate were sown and resultant M_3 seeds were harvested. NIRS analysis of seeds (M_3) from some of the single plants showed that a portion of the single plants kept the quality trait(s), while others did not. For example, Q3-5-23 (plant no. of 2011) was grown from an M_2 seed (seed no. of 2010: E1-3-343-8) with 57.78% oil, and 23 seeds of the plant were analyzed by NIRS, of which 18 contained higher than 55% oil, with the highest being 58.71%. Q3-12-24 (plant no. of 2011), grown from a seed with 62.97% oleate (seed no. of 2010: E1-2-11-16), still had 60.55% oleate in M_3 generation.

To summarize, the present study demonstrated the successful use of NIRS in selecting a limited number of peanut breeding materials with desired altered quality characters from large populations of M_2. Some of the single seeds with high oil, protein or oleate content lost the quality trait(s) in subsequent generation; these quality traits appeared to be caused by physiological abnormalities rather than genetic changes. However, there were other M_3 seeds with quality traits inherited from previous generation, which could be ascribed to mutations in related genes. We have reported a high oleate EMS mutant of LF 2 (2010 seed no.: E2-4-83-12), an output of the present study, that had dysfunctional mutated *FAD2A* (G448A in the coding region) and *FAD2B* (C313T in the coding region) [4]. Despite the uncertainty in quality traits, selection in M_2 seeds is still necessary as it helps to reduce the population size of M_2 plant and hence M_3 seed generations.

In a separate study, we identified several peanut induced/ natural mutants with over 70% oleate content, and used them in a hybridization program [3]. Thus far, a great number of high oleate lines including those with double "high"- high oleate and high yield, or high oleate and high shelling percentage, have been tentatively developed. Notably, some of the large seeded lines had a high shelling outturn when planted in Sanya, Hainan province (located in tropical zone), and some lines consistently exhibited high yield when planted in Laixi (located in temperate zone) and Sanya.

A Bold-Seeded Peanut Natural Mutant of *A. Duranensis* with High Oil Content

Breeding high oil peanut cultivars especially those with large seeds is most challenging, and variations in oil between years and locations may be quite high. In fact, in China, several high oil peanut cultivars have been released in Hebei, Hubei and Henan, but all of these peanut cultivars were found to only contain less than 54% seed oil when planted in Shandong peninsula. Availability of high oil genetic resources is of vital importance to the genetic improvement of the cultivated peanut in this region and the like. High oil peanut wild species have been reported by several research groups, but none of them possessed large seeds. A variant of a wild peanut species, *Arachis duranensis* PI 262133, was identified for the first time as with high oil content and large seeds at SPRI.

Cloning and sequencing of the rDNA ITS internal transcribed spacer (ITS) sequences from *A. duranensis* and the bold-seeded genotype detected no difference in their ITS sequences, suggesting that the variant was unlikely to be a result of natural hybridization.

The mutant was identified in a separate square cement block allotted to *A. duranensis* in the wild peanut nursery at SPRI Experiment Station, Laixi. The plant was grown from a seed of similar size to that of *A. duranensis*, but was found to possess thicker branches, and larger leaflets, pods and seeds (Figure 1, Table 9). The size and weight of pods and seeds of wild and mutant type *A. duranensis*significantly differed (P<0.01).

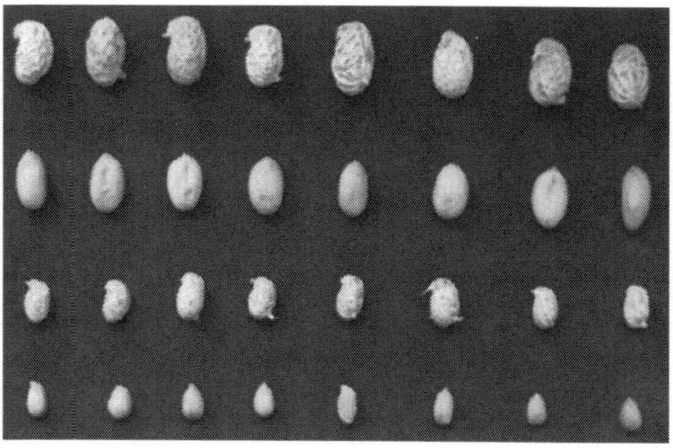

Figure 1. Pods and seeds of mutant type (top 2 rows) and wild type (bottom 2 rows) of *A. duranensis*

Table 9. Difference in pod and seed size and weight between wild and mutant type *A. duranensis*

Trait		Mean	SD
Pod weight (g)	Mutant type	0.93	0.24
	Wild type	0.24	0.05
Pod length (cm)	Mutant type	2.12	0.23
	Wild type	1.28	0.10
Pod thickness (cm)	Mutant type	1.30	0.16
	Wild type	0.75	0.06
Pod width (cm)	Mutant type	1.11	0.12
	Wild type	0.70	0.07
Seed weight (g)	Mutant type	0.67	0.17
	Wild type	0.19	0.05

Seed length (cm)	Mutant type	1.67	0.20
	Wild type	1.03	0.09
Seed thickness (cm)	Mutant type	0.92	0.13
	Wild type	0.57	0.07
Seed width (cm)	Mutant type	0.70	0.08
	Wild type	0.55	0.06

Totally 110 seeds were analyzed by NIRS to predict their oil content (Figure 2). Of them, 40 contained more than 55% oil, with 60.33% being the highest. The average oil content of the mutant was 54.23%, as against 52.62% in the wild type *A. duranensis*.

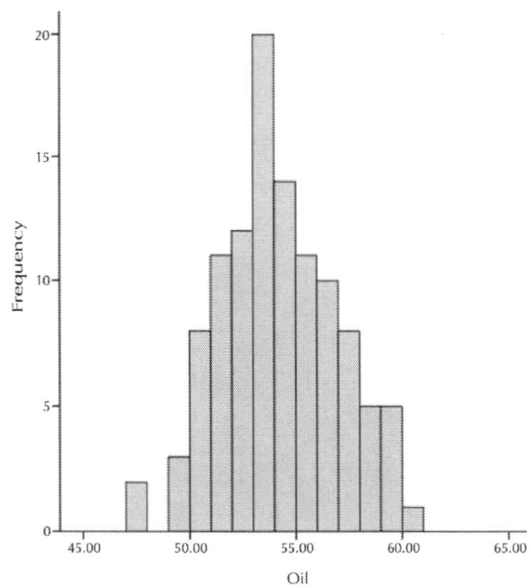

Figure 2. Frequency distribution of seed oil content (%) in 110 seeds of the *A. duranensis* mutant

According to latest classification proposed by Krapovickas and Gregory (1994), the genus *Arachis* is devided into 9 sections, consisting of more than 80 species [19]. Of them, *A. hypogaea* L. is widely cultivated for oil extraction and food uses. Other *Arachis* species of economic importance include *A. glabrata, A. pintoi,* and *A. duranensis,* which are being utilized as forage and/or ground cover [20]. Wild species have higher genetic diversity than the cultivated peanut, providing a desirable source for stress resistance, high oil, protein, oleate, amino acid content as well as high yield factors [20].

Whether the mutant was resulting from chromosome doubling or gene mutation is still unknown. Chromosome counting is absolutely necessary. Anyhow, the mutant reported here is of relevance both to the genetic improvement of the cultivated peanut and to its direct utilization as forage and/or ground cover.

CONCLUSION

Our study on peanut mutagenesis clearly demonstrated that it was possible to induce mutations in productivity, oleate, oil and protein content in the cultivated peanut. NIRS may facilitate identification of induced/natural mutants with improved quality characters. Although not all of the variations in quality traits of M_2 seeds are inheritable, selection in this generation may help to reduce the size of M_3seed population.

ACKNOWLEDGEMENT

The authors thank the financial support from China Agricultural Research System (Grant No. CARS-14), Ministry of Agriculture, China, Qingdao Science & Technology Support Program (Grant No. 10-3-3-20-nsh, Grant No. 09-1-3-67-jch), and Jilin Natural Science Foundation (Grant No. 20101580).).

REFERENCES

1. McGill JF.Economic importance of peanuts. In: Peanut-Culture and uses, a symposium. American Peanut Research and Education Association. 1973316
2. V. R. Rao, . Chapter, In. Botany, P. S. Reddy, (ed, Groundnut, Indian Council of Agricultural Research (ICAR). 19882464
3. C. T. Wang, D. Q. Hu, Y. F. Ding, H. T. Yu, Y. Y. Tang, X. Z. Wang, J. C. Zhang, D. X. A. Chen, set. new, allelespecific. P. C. R. of, for. primers, of. identification, hybrids. true, normal. in, ×. oleate,

oleate. high, in. crosses, groundnut, Journal of SAT Agricultural Research 2011ejournal.icrisat.org/9Groundnut/A_new_set.pdf (accessed 6 June 2012)

4. Fang CQ, Wang CT, Wang PW, Tang YY, Wang XZ, Cui FG, Yu SL.Identification of a novel mutation in FAD2B from a peanut EMS mutant with elevated oleate content. Journal of Oleo Science 2012613143148
5. Hammons RO. Peanuts: genetic vulnerability and breeding strategy.Crop Science 1975194527530
6. Wang CT, Yang XD, Xu JZ, Liu GZ.Differences in pod characters among L7-1 peanut cultivar and its chemical mutants. International Arachis Newsletter 2006261214
7. Zhang JC, Wang CT, Tang YY, Wang XZ.Effects of grading on the main quality attributes of peanut kernels. Frontiers of Agriculture in China 200933291293
8. G. P. Fox, A. Cruickshank, Near infrared reflectance as a rapid and inexpensive surrogate measure for fatty acid composition and oil content in peanuts (Arachis hypogaea L.). Journal of Near Infrared Spectroscopy 20055287291
9. S. L. Yu, Y. J. Zhu, P. Min, H. Liu, Y. L. N. I. R. S. Cao, determine. to, protein, contents. oil, peanut. in, kernels, Journal of Peanut Science 2003Supplement):138143
10. S. L. Yu, Y. J. Zhu, P. Min, H. Liu, Y. L. Cao, C. T. Wang, C. S. Zhang, X. Liu, X. Q. N. I. R. S. Zhou, determine. to, fatty. major, contents. acids, peanut. in, kernels, In: Wan SB. (ed.) High Quality and High Output Peanuts: Principle and techniques for Production. Beijing: China Agricultural Science & Technology Press; 2003344349
11. B. Tillman, D. Gorbet, G. Person, Predicting oleic and linoleic acid content of single peanut seeds using near-infrared reflectance spectroscopy. Crop Science 2006521212126
12. J. Sundaram, C. V. Kandala, C. L. Butts, Application of near infrared spectroscopy to peanut grading and quality analysis: overview. Sensing and Instrumentation for Food Quality and Safety 20093156164
13. Wang CT, Yu SL, Zhang SW, Wang XZ, Tang YY, Zhang JC, Chen DX.Novel protocol to identify true hybrids in normal oleate x

high oleate crosses in peanut. Electronic Journal of Biotechnology 2010http://dx.doi.org/10.2225/5-fulltext -18accessed 6 June 2012)

14. Wang CT, Tang YY, Wang XZ, Zhang SW, Li GJ, Zhang JC, Yu SL.Sodium azide mutagenesis resulted in a peanut plant with elevated oleate content. Electronic Journal of Biotechnology. 2011http://dx.doi.org/10.2225/2-fulltext -4accessed 6 June 2012)
15. Wang CT, Wang XZ, Tang YY, Chen DX, Zhang JC, Cui FG, Yu SL.High yielding mutants achieved by injecting EMS into peanut flower organs. Journal of Nuclear Agricultural Sciences 2010242239242
16. Wang CT, Wang XZ, Tang YY, Zhang JC, Chen DX, Xu JZ, Yang XD, Song GS, Cui FG. Huayu 40, a groundnut cultivar developed through EMS mutagenesis.Journal of SAT Agricultural Research 2011ejournal.icrisat.org/9Groundnut/Huayu40.pdf (accessed 6 June 2012)
17. J. Zheng, Z. Zhang, L. Chen, Q. Wan, M. Hu, W. Wang, K. Zhang, D. Liu, X. Chen, X. I. T. Wei-I, S. J. marker, application. its, construction. in, upland. of, linkage. cotton, map, Scientia Agricultura Sinica 200841822412248
18. S. B. Wan, editor, Peanut cultivation Science in China. Shanghai: Shanghai Science and Technology Press. 2003pp.
19. A. Krapovikas, W. C. Gregory, Taxonomia del genero Arachis (Leguminosae). Bonplandia 19948186
20. Yu SL, Wang CT, Yang QL, Zhang DX, Zhang XY, Cao YL, Liang XQ, Liao BS, editors.Peanut Genetics and Breeding in China. Shanghai: Shanghai Science and Technology Press; 2011pp.

Chapter 5

Arachis Batizocoi: A Study of its Relationship to Cultivated Peanut (A. hypogaea) and its Potential for Introgression of wild genes into the Peanut crop using Induced Allotetraploids

Soraya C. M. Leal-Bertioli[1], Silvio P. Santos[2,3], Karinne M. Dantas[1,3], Peter W. Inglis[1], Stephan Nielen[1,4], Ana C. G. Araujo[1], Joseane P. Silva[1], Uiara Cavalcante[1,2], Patricia M. Guimarães[1], Ana Cristina M. Brasileiro[1], Noelia Carrasquilla-Garcia[5], R. Varma Penmetsa[5], Douglas Cook[5], Márcio C. Moretzsohn[1] and David J. Bertioli[2,3]

[1]Embrapa Genetic Resources and Biotechnology, PqEB W5 Norte Final, CP 02372, CEP 70.770-917, Brasília, DF, Brazil,

[2]University of Brasilia, Institute of Biological Sciences, Campus Darcy Ribeiro, CEP 70.910-900, Brasília, DF, Brazil,

[3]Catholic University of Brasilia, Biotechnology and Genomic Sciences, SGAN 916 Avenida W5, CEP 70.790-160, Brasilia, DF, Brazil,

[4]Plant Breeding and Genetics Section, Joint FAO/IAEA Division, International Atomic Energy Agency, Vienna International Centre, Vienna A-1400, Austria and

[5]Department of Plant Pathology, University of California at Davis, One Shields Avenue, Davis, CA 95616, USA

ABSTRACT

Background and Aims

Arachis batizocoi is a wild relative of cultivated peanut (*A. hypogaea*), an allotetraploid with an AABB genome. *Arachis batizocoi* was once considered the ancestral donor of the peanut B genome, but cytogenetics and DNA phylogenies have indicated a new genome classification, 'K'. These observations seem inconsistent with genetic studies and breeding that have shown that *A. batizocoi* can behave as a B genome.

Methods

The genetic behaviour, genome composition and phylogenetic position of *A. batizocoi* were studied using controlled hybridizations, induced tetraploidy, whole-genome *in situ* fluorescent hybridization (GISH) and molecular phylogenetics.

Key Results

Sterile diploid hybrids containing AK genomes were obtained using *A. batizocoi* and the A genome species *A. duranensis*, *A. stenosperma*, *A. correntina* or *A. villosa*. From these, three types of AAKK allotetraploids were obtained, each in multiple independent polyploidy events. Induced allotetraploids were vigorous and fertile, and were hybridized to *A. hypogaea* to produce F_1 hybrids. Even with the same parental combination, fertility of these F_1 hybrids varied greatly, suggesting the influence of stochastic genetic or epigenetic events. Interestingly, hybrids with *A. hypogaea* ssp. *hypogaea* were significantly more fertile than those with the subspecies *fastigiata*. GISH in cultivated × induced allotetraploids hybrids (harbouring AABK genomes) and a molecular phylogeny using 16 intron sequences showed that the K genome is distinct, but more closely related to the B than to the A genome.

Conclusions

The K genome of *A. batizocoi* is more related to B than to the A genome, but is distinct. As such, when incorporated in an induced allotetraploid (AAKK) it can behave as a B genome in crosses with peanut. However, the fertility of hybrids and their progeny depends upon the compatibility of the A genome interactions. The genetic distinctness of *A. batizocoi* makes it an important source of allelic diversity in itself, especially in crosses involving *A. hypogaea* ssp. *hypogaea*.

INTRODUCTION

The genus *Arachis* is endemic to South America, found over a wide area and in diverse environments. It contains 81 described species, assembled into nine taxonomical sections, according to their morphology, geographic distribution and cross-compatibility (Krapovickas and Gregory, 1994; Valls and Simpson, 2005; Valls *et al.*, 2013). Peanut and another 30 known wild species reside in the Section *Arachis*. Most species are diploid ($2n=2x=20$), three are aneuploid or dysploid ($2n=2x=18$) and two are allotetraploids ($2n=4x=40$). The allotetraploids, the wild *A. monticola*and the cultivated peanut *A. hypogaea* are very closely related and belong to the same biological species. Within and between these tetraploid species, DNA polymorphism is very low (Halward *et al.*, 1991; Kochert*et al.*, 1996), and it seems most likely that a single, or very few polyploidy events involving the same ancestors, gave rise to both species only a few thousand years ago (Bertioli *et al.*, 2011).

Peanut is a very important food crop throughout the tropics and sub-tropics, and its very narrow genetic base presents limitations for crop improvement. These limitations have long been apparent for breeding for pest and disease resistances, and there are strong reasons to believe that introducing allelic diversity from wild species could also improve more complex traits such as yield and drought response. Correspondingly, peanut breeders have been interested in introducing wild species into breeding programmes. However, the ploidy level difference and fertility barriers between peanut and its wild relatives hamper their effective use. Clearly, an understanding of the relationships of the cultivated and wild species would facilitate the production of fertile hybrids and, consequently, introgression of wild genes into the peanut crop. In this regard, many studies have focused on the species with 20, mostly metacentric chromosomes in the section *Arachis*, because these include cultivated peanut and its closest wild relatives (here we will not deal with the $2n=2x=18$ or asymmetric karyotype species). These species were initially divided into two genome types, A and B, based on the component sub-genomes of *A. hypogaea*. The A genomes are characterized by the presence of a small pair of chromosomes with allocyclic condensation, 'the A chromosomes' (Husted, 1936; Smartt *et al.*, 1978), whereas the B genomes are characterized by the absence of these A chromosomes. The validity of the A genome grouping is strongly supported by diverse evidence. This includes their moderate to high cross-compatibility; unifying biological characteristics, such as predominantly perennial growth and chromosomes with condensed A/T-rich centromeres; and their grouping in molecular phylogenies. The definition of the B genome group is more precarious since it is based on the absence of a character (the small A chromosomes). Accordingly, further cytogenetic and molecular phylogenetic studies justified the sub-division of species traditionally assigned to this group: *A. ipaënsis*, which is now thought to be ancestral to *A. hypogaea* and its closest relatives remains the B genome [also known as the B genome *sensu stricto* (*s.s.*)], while *A. batizocoi* and two closely related species are assigned as a new grouping, the 'K' genome, and another two species are assigned to the F genome (Robledo and Seijo, 2010). The data available on chromosome behaviour during meiosis in diploid hybrids are consistent with the K genome being distinct from both the A and B genomes. Intragenome hybrids (A × A, B × B and K × K) generally show a mean bivalent formation of the chromosomes >9·5 out of the expected 10, whilst intergenome hybrids (A × B, A × K and B × K) show values <8 (Robledo and Seijo, 2010). Furthermore, with this revision and subsequent works, the phylogenetic position of the K genome relative to the A and B genomes in *Arachis* became a matter of uncertainty. A notable feature that distinguishes the K genome chromosomes is the presence of an A genome-like condensed centromere, but the absence of the 'A chromosome'. Also, whilst some molecular phylogenetic studies placed the species assigned as K genome closer to the B genome, (Moretzsohn *et al.*, 2004; Bechara *et al.*, 2010; Friend *et al.*, 2010), others placed it closer to the A genome (Tallury *et al.*, 2005; Robledo *et al.*, 2009;Robledo and Seijo, 2010; Moretzsohn *et al.*, 2013).

The establishment of the K genome group sits uncomfortably with the results from a series of classic experiments, which were started when *A. batizocoi* was considered the B genome donor to the cultivated species (Smartt *et al.*, 1978; Stalker *et al.*, 1991). Using a hybridization scheme known as 'the tetraploid route', Simpson *et al.* (1993) obtained a wild-derived allotetraploid (also referred to as amphidiploid). To do this, an A genome diploid (in this case, a hybrid between *A. cardenasii* and *A. diogoi*) was crossed with *A. batizocoi*, giving origin to a sterile hybrid. This was treated with colchicine to induce chromosome doubling and restore fertility. The resultant allotetraploid [*A. batizocoi* × (*A. cardenasii* × *A. diogoi*)]$^{4\times}$ was registered as TxAG-6 and produced viable hybrids when crossed with cultivated peanut. A BC_1 population derived from the peanut cultivar Florunner and TxAG-6 was used to generate the first peanut tetraploid map (Burow *et al.*, 2001). This map showed that the genome of*A. batizocoi* recombined with the B genome of *A. hypogaea* in a relatively regular mode. Furthermore, successive backcrossing followed by selection allowed the development of a number of important cultivars such as COAN, NemaTAM and Tifguard (Simpson and Starr, 2001; Simpson *et al.*, 2003; Holbrook *et al.*, 2008), which harbour a chromosome segment from*A. cardenasii* conferring root-knot nematode resistance (Nagy *et al.*, 2010).

Here we used controlled hybridizations, induced polyploidy, whole-genome *in situ* fluorescent hybridization (GISH) and the construction of a molecular phylogeny to better define the K genome of *A. batizocoi* and its potential for use in hybridization schemes designed to introduce wild alleles into the peanut crop.

MATERIALS AND METHODS

Production of Hybrids and Induced Allotetraploids

Arachis seeds were obtained from the Active Germplasm Bank of Embrapa Genetic Resources and Biotechnology (Cenargen, Brasília, Brazil). Plants were grown in an open-plan greenhouse. Diploid wild genotypes used for crosses are listed in Table 1. *Arachis batizocoi* was used as female parent and A genome accessions were used as male parents. All crosses were performed essentially as described by Simpson (1991), between the months of October and February.

Hybrid plants were initially identified by their aggregated pollen masses and absence of peg production. To confirm hybrid identity, marker analysis with the simple sequence repeat (SSR) Ah-041 (Moretzsohn *et al.*, 2004) was used. Genomic DNA was extracted from fresh, young leaves according to Ferreira and Grattapaglia (1998). PCRs were performed in a 13 µL volume: 1× PCR buffer (10 mM Tris–HCl pH 8·3, 50 mM KCl, 2·0 mM$MgCl_2$), 0·2 mM of each dNTP, 1 U of *Taq* DNA polymerase (Invitrogen, USA), 1 ng µL^{-1} bovine serum albumin (BSA), 5 pmol of primer (Ah-041-Fwd, 5 -CGCCACAAGATTAACAAGCACC-3 ; Ah-041-Rev, 5 -GCTGGGATCATTGTAGGGAAGG-3) and 10 ng of genomic DNA. Amplifications were done in an ABI 9700 thermocycler (Applied Biosystems, Foster City, CA, USA) with the following cycling conditions: 96 °C for 2 min, followed by 30 cycles of 94 °C for 1 min, 60 °C for 1 min, 72 °C for 1 min and a final extension of 72 °C for 7 min. Fragments were separated in a 5 % polyacrylamide gel, and stained with silver nitrate (Creste *et al.*, 2001).

Table 1. Wild and cultivated *Arachis*genotypes used for crosses; induced allotetraploids andF_1 hybrids were produced in this study

Genotype	Plant ID	**Ploidy**	**Genome type**	**Collection site/ classification**
Wild diploid species				
A. batizocoi Krapov. & W.C. Gregory	K9484	2*x*	KK	Parape, Bolívia
A. batizocoi Krapov. & W.C. Gregory	GKBSPSc30081*	2*x*	KK	Lagunillas, Bolívia
A. correntina (Burkart) Krapov. & W.C. Gregory	GKP9548	2*x*	AA	Corrientes, Argentina
A. duranensis Krapov. & W.C. Gregory	V14167	2*x*	AA	Salta, Argentina
A. duranensis Krapov. & W.C. Gregory	SeSn2848	2*x*	AA	Salta, Argentina
A. duranensis Krapov. & W.C. Gregory	KSSc3603*	2*x*	AA	Salta, Argentina
A. duranensis Krapov. & W.C. Gregory	WiSVg1510-B*	2*x*	AA	Boquerón, Paraguay
A. ipaënsis Krapov. & W.C. Gregory	KG30076	2*x*	BB	Gran Chaco, Bolivia
A. stenosperma Krapov. & W.C. Gregory	V13828	2*x*	AA	Luis Alves, GO, Brazil
A. stenosperma Krapov. & W.C. Gregory	V10309	2*x*	AA	Rondonopolis, MT, Brazil
A. stenosperma Krapov. & W.C. Gregory	V7762	2*x*	AA	Araguaiana, MT, Brazil
A. villosa Benth	V12812	2*x*	AA	Bella Unión, Uruguay
Peanut accessions				
A. hypogaea subsp. *fastigiata* var. *fastigiata* L.	BR1	4*x*	AABB	Modern cultivar

A. hypogaea subsp. *fastigiata* var. *fastigiata* L.	55437 (Senegal)	4*x*	AABB	Modern cultivar
A. hypogaea subsp. *fastigiata* var. *fastigiata* L.	BRS-Havana	4*x*	AABB	Modern cultivar
A. hypogaea subsp. *hypogaea* var. *hypogaea* L.	IAC-Caiapó	4*x*	AABB	Modern cultivar
A. hypogaea subsp. *hypogaea* var. *hypogaea* L.	Runner IAC 886	4*x*	AABB	Modern cultivar
Induced allotetraploids				
[*A. batizocoi* K9484 × *A. duranensis* V14167]$^{4\times}$	BatDur1	4*x*	AAKK	Synthetic alotetraploid
[*A. batizocoi* K9484 × *A. duranensis* SeSn2848]$^{4\times}$	BatDur2	4*x*	AAKK	Synthetic alotetraploid
[*A. batizocoi* K9484 × *A. stenosperma* V10309]$^{4\times}$	BatSten1	4*x*	AAKK	Synthetic alotetraploid
F_1 hybrids between induced allotetraploids and peanut				
IAC-Caiapó × [*A. batizocoi* K9484 × *A. duranensis* V14167]$^{4\times}$	Caiapo × BatDur1	4*x*	AABK	F_1 hybrid
IAC-Caiapó × [*A. batizocoi* K9484 × *A. duranensis* SeSn2848]$^{4\times}$	Caiapo × BatDur2	4*x*	AABK	F_1 hybrid
IAC-Caiapó × [*A. batizocoi* K9484 × *A. stenosperma* V10309]$^{4\times}$	Caiapo × BatSten1	4*x*	AABK	F_1 hybrid
Runner IAC 886 × [*A. batizocoi* K9484 × *A. duranensis* V14167]$^{4\times}$	Runner × BatDur1	4*x*	AABK	F_1 hybrid

Runner IAC 886 × [*A. batizocoi* K9484 × *A. duranensis* SeSn2848]$^{4\times}$	Runner × BatDur2	4*x*	AABK	F_1 hybrid
Runner IAC 886 × [*A. batizocoi* K9484 × *A. stenosperma* V10309]$^{4\times}$	Runner × BatSten1	4*x*	AABK	F_1 hybrid
Runner IAC 886 × [*A. ipaënsis* KG30076 × *A. duranensis* V14167]$^{4\times}$	Runner × IpaDur1	4*x*	AABB	F_1 hybrid
BR1 × [*A. batizocoi* K9484 × *A. duranensis* V14167]$^{4\times}$	BR1 × BatDur1	4*x*	AABK	F_1 hybrid
BR1 × [*A. batizocoi* K9484 × *A. duranensis* SeSn2848]$^{4\times}$	BR1 × BatDur2	4*x*	AABK	F_1 hybrid
BR1 × [*A. batizocoi* K9484 × *A. stenosperma* V10309]$^{4\times}$	BR1 × BatSten1	4*x*	AABK	F_1 hybrid
BRS Havana × [*A. batizocoi* K9484 × *A. duranensis* V14167]$^{4\times}$	Havana × BatDur1	4*x*	AABK	F_1 hybrid
BRS Havana × [*A. batizocoi* K9484 × *A. duranensis* SeSn2848]$^{4\times}$	Havana × BatDur2	4*x*	AABK	F_1 hybrid
BRS Havana × [*A. batizocoi* K9484 × *A. stenosperma* V10309]$^{4\times}$	Havana × BatSten1	4*x*	AABK	F_1 hybrid
55-437 × [*A. batizocoi* K9484 × *A. duranensis* V14167]$^{4\times}$	55-437 × BatDur1	4*x*	AABK	F_1 hybrid
55-437 × [*A. batizocoi* K9484 × *A. duranensis* SeSn2848]$^{4\times}$	55-437 × BatDur2	4*x*	AABK	F_1 hybrid
55-437 × [*A. batizocoi* K9484 × *A. stenosperma* V10309]$^{4\times}$	55-437 × BatSten1	4*x*	AABK	F_1 hybrid

Confirmed hybrid plants (harbouring AK genomes) were treated with colchicine to induce chromosome doubling (Simpson, 1991). Cuttings from lateral branches were submerged in a solution of 0·2 % colchicine for 10 h. The cuttings were then rinsed in water for 15 min and planted in sand. Upon rooting and the appearance of new shoots, cuttings were transferred to a substrate containing 60 % sand and 40 % red clay-rich soil supplemented by 1500 g of NPK 4-30-16 per m^3 of soil. The production of pegs and seeds signalled the recovery of fertility of some sectors in the colchicine-treated hybrids. Progeny from these treated hybrids were considered putative induced allotetraploids. Cytogenetic analysis (see below) was used to confirm their ploidy level and chromosome complement.

Morphological Measurements of Induced Allotetraploids and Diploid Parentals

Dimensions of seeds of allotetraploids and their diploid parents were determined using a digital caliper, at the longest and widest point of each seed. Ten fresh seeds/plant were measured and weighed. For plants that produced <10 seeds, all seeds available were evaluated. The first fully expanded leaf from the apex of at least eight different plants of each genotype were scanned and the areas were determined using the software QUANT® (Vale *et al.*, 2001). *Arachis hypogaea* cv. Tatu was included as control.

Hybridization of *A. hypogaea* with Induced Allotetraploids

Five cultivars belonging to the two subspecies of peanut, *hypogaea* and *fastigiata*, were used as female parents in crosses with the obtained allotetraploids (Table 1). Crosses were performed between the months of February and June. To determine crossability of allotetraploids with cultivated peanut genotypes (the number of F_1 hybrids produced per cross), pollinations and peg production were logged daily. Putative F_1 hybrids were confirmed by the presence of yellow flower colour inherited from the allotetraploids used as male parents (the five cultivars used here have orange flowers). Pollen viability of F_1 hybrids was estimated by the staining method with acetic carmine (Linsley and

Cazier, 1963). For each genotype, 1000 pollen grains were analysed from oblong anthers as follows: 100 pollen grains per anther, two anthers per flower and five flowers per hybrid. Hybrid fertility was estimated by observation of peg production and by the number of seeds produced per F_1 hybrid.

Statistical Analyses

All data were analysed using the statistical software R. To verify normality, the Shapiro–Wilk test was used. When data presented normality, the comparisons were performed using standard variance analyses, followed by the Tukey test of multiple comparisons. When data did not follow normal distribution, generalized linear models were used to relax assumptions about response variable and errors. Significance level in all analyses was 0·05. Average silhouette width, $s(i)$, using K-means was used to create groupings (values vary between 0 and 1, and observations closer to 1 were better clustered).

Cytogenetic Analyses

Clean young root tips for cytogenetic analysis were produced by rooting the petioles of leaves in moist cotton wool in humid Petri dishes at about 25 °C in illuminated conditions. Root tips were treated with 2 mM 8-hydroxyquinoline for 2 h at room temperature and another hour on ice prior to fixation in ethanol: macetic acid solution (3:1, v/v) for 24 h at 4 °C. Fixed root tips were incubated in enzyme buffer (4 mM citric acid, 6 mMsodium citrate, pH 4·8) containing 2 % cellulose Onozuka R10 (Merck) and 20 % pectinase (Sigma) for 45 min at 37 °C. Chromosome spreads were made using a modified technique originally described by Pijnacker and Ferwerda (1984). Briefly, meristem cells of a single root tip were suspended in 15 µL of 60 % acetic acid on a microscope slide. The slide was heated to 45 °C, another drop of acetic acid was added and, after 1 min, the chromosome suspension was precipitated by adding fixative solution in excess.

Genomic DNAs for GISH consisted of sheared genomic DNA, labelled with either digoxigenin-11-dUTP or biotin-11-dUTP (Roche) by random primed labelling. Here, digoxigenin-labelled genomic DNA of *A. duranensis* and biotin-labelled genomic DNA of *A.*

ipaënsis were used, both at 65 ng µL^{-1}. Pre-treatment, hybridization, washing and detection procedures essentially followed the protocols of Schwarzacher and Heslop-Harrison (2000). Denatured DNA probes were mixed in a hybridization solution containing 50 % (v/v) formamide, 10 % (w/v) dextran sulphate, 0·125 mMEDTA, 0·125 % (w/v) SDS and 1 µg of salmon sperm DNA. The chromosomes and hybridization mix were denatured together at 81·5 °C for 10 min before hybridization overnight at 37 °C. Post-hybridization washes were carried out at 83 % stringency. Hybridization sites were detected by sheep anti-digoxigenin conjugated to fluorescein isothiocyanate (antibody anti-dig conjugated to FITC; green fluorescence) and Cy3–streptavidin (red fluorescence). FITC signals were amplified using anti-sheep antibody conjugated to fluorescein (Vector Laboratories, USA). Chromosomes were counterstained with DAPI (4′,6-diamidino-2-phenylindole; blue fluorescence). Slides were observed in a Zeiss Axiophot epifluorescence microscope using appropriate filters. Images were captured and analysed using the CCD camera AxioCam MRm in combination with the AxioVison software (both Zeiss, Germany) and processed in Adobe Photoshop CS.

Phlyogenetic Analysis of A, B and K Genomes

Intron Sequencing

In order to better define the phylogenetic position of *A. batizocoi* with respect to the *Arachis* A and B *s.s.* genomes, we used orthologous sequences from three accessions of *A. duranensis* (A genome; accession nos K7988, KSSc3603 and WiSVg1510-B), one of *A. stenosperma* (A genome; V10309), two of *A. batizocoi* (K genome, *sensu lato* B genome; K9484 and GKBSPSc30081), one of *A. ipaënsis* (B genome *s.s.*; KG30076) and one from *A. magna* (B genome *s.s.*; KG30097). Orthologous coding regions tend to be highly conserved and thus phylogenetically uninformative in closely related species. Therefore we used orthologous intron sequences, amplifying them via PCR primers complementary to conserved exon sequences that flank the introns (Choi *et al.*, 2006; Penmetsa and Cook, unpubl. res.). Primer pairs designed to amplify the 16 tentative orthologous genes (TOGs) are described in Table 2. Sequencing was performed using the

BigDye 3.1 Terminator sequencing kit (Applied Biosystems) on ABI sequencers (Applied Biosystems). Sequences were processed using the Staden Package (Staden, 1996), with base calling using Phred (Staden, 1996; Ewing and Green, 1998).

Table 2. Primer pairs used for amplication of Sanger-sequence tentative orthologous genes (TOGs)

TOG amplicon	**F oligo sequence (5 >3)**	**R oligo sequence (5 >3)**
TOG896852	GGYTCCTTGCAAATGTTCRT	GAGGCCAATAAAGCCATTCA
TOG896966	GTGCTCTTCCCATTTCRGTA	TCCATGGTTTTGCTTCATCA
TOG897172	AGCATATCATTRGAGCCWGG	CACTTTGGCTGGAAACCAAT
TOG897242	GATGGGATCCSAAGCAACTW	GRGTATGCCCAAGAAGTTCM
TOG897648	CGATSYTGTGGGAGAACATT	WGGAGCTTCACAAATGCAAA
TOG898975	AARGCTCTTGRCACCTTTCA	GGACCGGGACAGACMACRGT
TOG899382	GTCTTGGAGAAGACCGCTTG	CATCATGCACAAGTTCCTKG
TOG899751	GTTGACCTGCACCCTYTGYT	TTCCATTGCAATAGCATCCA
TOG900140	GGCMCATGTTGTTATGGCTG	TTCCAWGCTTCAGCAAACCT
TOG900220	CATTTGCWTTAGCTGCCTCA	CCATGCTGAGWACTTGCGAA
TOG901841	WTGCATCATTGGGSAATCTA	AAGGTCCCATATGTTYGCAS
TOG903928	GACTGTTCGCCAYCATGCTR	CCTTTGCTRCTTTCCCACCR
TOG904180	CGTGCCAAAGCTGTRCTTSA	GGTGTTCAAGAARCGACCMC
TOG908826	TCCRGGTGAAGGTTGTTAAG	CTGGTTGTCTGCAGATTGGA
TOG910656	GATTGCATCCTGATGGTYCT	CAACACGGGAGCCTACRGAT
TOG923111	TAGCAGTGGTGGTGGTCAAA	CWGGYCCACCAAAGAGATAA

Sequence Alignment and Phylogenetic Analysis

Arachis intron sequences were aligned using Muscle (Edgar, 2004), and concatenated in SequenceMatrix 1.7.8 (Vaidya *et al.*, 2011). Several of the TOG introns were found to be rich in indels, which were encoded and appended to the data matrix as C or A base transversions for presence or absence, respectively, using the simple coding scheme of Simmons and Ochoterena (2000), as implemented in the program Fastgap 1.2 (Borchsenius, F. 2009. FastGap 1.2. Department of

Biosciences, Aarhus University, Denmark. Published online athttp://www.aubot.dk/FastGap_home.htm).

A phylogenetic hypothesis based on the concatenated TOGs and encoded indel characters was obtained using the Bayesian Markov Chain Monte Carlo (MCMC) method as implemented in MrBayes 3.2.2 (Ronquist *et al.*, 2012). Reversible-jump MCMC (Huelsenbeck *et al.*, 2004) was invoked for model optimization within the GTR family (nst=mixed rates=invgamma), and priors allowed to vary independently for each of the 17 data partitions (each TOG+indel matrix). One cold and three heated MCMC chains were run in parallel for 2 million generations, sampling every 1000 generations. This run time was sufficient for the convergence diagnostic, the mean standard deviation of split frequencies, to fall to 0·003612. The first 25 % of the trees were discarded (the burn-in) prior to calculation of the 50 % majority rule consensus tree. Convergence and adequacy of the burn-in were confirmed by visualization of tree probabilities in the program Tracer v1.5 (Rambaut *et al.*, 2014), available fromhttp://beast.bio.ed.ac.uk/Tracer, where a normal distribution of log likelihood values was obtained for both MCMC runs (data not shown).

Testing for Ancient Hybridization Events in the Arachis Species Evolution

To detect possible ancient hybridization events and reticulate evolution between the genomes of the included *Arachis* species, we analysed the aligned and concatenated matrix of the 16 TOGs and appended coded indel characters in SplitsTree 4 (Huson and Bryant, 2006), using the parsimony splits method. We also compared the results with a similar analysis where data from *A. hypogaea* 'Runner IAC 886' were included in the matrix as a recombination-positive control. The PHI statistical test (Bruen *et al.*, 2006) was also used to infer recombination along the concatenated alignment, with a window size of 100 bases. We also attempted to detect recombination using the Bayesian dual Multiple Change-Point (MCP) model, with a default starting window size of 200 bases, as implemented in the DualBrothers program (Minin *et al.*, 2005) within the Geneious environment (Drummond *et al.*, 2011).

RESULTS

Production of Hybrids and Artificially Induced Allotetraploids

Sixty-seven marker-confirmed diploid AK genome hybrid plants, representing seven different hybrid combinations, were generated with crosses between *A. batizocoi* and seven accessions containing the A genome of four different species (Table 3). All diploid hybrids had aggregated pollen masses and did not produce progeny. Five out of the seven combinations were vigorous, but two, *A. batizocoi* × *A. correntina*and *A. batizocoi* × *A. villosa*, showed altered morphological development. The former developed spontaneous tumour-like growth at the base of the stems, but otherwise grew vigorously. The latter developed spontaneous necrotic patches on the leaves, followed, when the plants grew older, by leaf drop and eventual death. These irregular phenotypes, indicative of hybrid incompatibilities, occurred with all independently obtained individuals of these hybrid combinations.

Table 3. Production of allotetraploid plants

Cross	N-hyb	N-allo	T/D
A. batizocoi K9484 × *A. correntina* GKP9548	18	0	0·00
A. batizocoi K9484 × *A. duranensis* V14167	12	23	1·92
A. batizocoi K9484 × *A. duranensis* SeSn2848	8	12	1·50
A. batizocoi K9484 × *A. stenosperma* V10309	9	31	3·44
A. batizocoi K9484 × *A. stenosperma* V13828	9	0	0·00
A. batizocoi K9484 × *A. stenosperma* V7762	7	0	0·00
A. batizocoi K9484 × *A. villosa* V12812	4	0	0·00
Total	67	66	

N-hyb, number of diploid hybrids; N-allo, number of induced allotetraploids obtained after colchicine treatment; T/D, allotetraploid/diploid hybrid ratio.

After treatment with colchicine, the stem cuttings showed some necrosis, delayed rooting and production of distorted leaves. Many died, but others successfully produced new plants. Of these, restoration of fertility was observed in three of the seven hybrid combinations. This was indicated by the production of gynophores and development of seeds. From these primary seeds, 66 independent fertile allotetraploid plants were produced. These induced allotetraploids plants combined the K genome of *A. batizocoi* K9484 with the A genome of *A. duranensis* V14167, *A. duranensis* SeSn2848 or *A. stenosperma* V10309. They were named BatDur1, BatDur2 and BatSten1, respectively (Table 3). Germinated allotetraploid seedlings usually presented developmental aberrations, such as the absence of chlorophyll in some leaflet areas, small curled distorted leaves and delimited sectors of necrosis in the young leaves (Fig. 1), but these characteristics disappeared about 3 weeks after germination. These characters reappeared in seedlings of subsequent generations, with an overall tendency of decreasing with each passing generation. Allotetraploid plants which originated from the same diploid parents, but from different polyploidy events, were phenotypically uniform, except for one white-flowered mutant of BatSten1.

Figure 1. A leaf of [*A. batizocoi* K9484 × *A. duranensis* SeSn2848]$^{4\times}$ (BatDur2) as an example of the phenotypic abnormalities typical of young leaves of induced allotetraploids, especially in early generations. Note the bleached and distorted leaf lamina and necrotic spots that appear to be spontaneous (A) and a plantlet with heavily bleached leaves (B).

Morphological Measurements of Induced Allotetraploids and Diploid Parentals

Diploid plants tend to have smaller leaves than tetraploid peanut. *Arachis batizocoi* is an exception, having leaves larger than any cultivated peanut measured here (Fig. 2). All adult induced allotetraploid plants, like their diploid parents, had trailing growth habit, but with larger overall size, larger leaves (Fig. 2) and greater vigour. The allotetraploids produced seed abundantly (ranging between 40·6 and 68·2 seeds per plant on average), in numbers comparable with those of their parental wild species (45·8–73·9), and with those of cultivated peanut (44·2–66·5), and did not differ among themselves. The likelihood ratio determined through generalized linear model Poisson was 7·457 and the *P*-value was 0·024. For the induced allotetraploids, weight, length and width of the seeds produced were slightly larger than those of wild parents, and lower than those of cultivated peanut (Kruskal–Wallis, $P<0{\cdot}05$). The exception is the allotetraploid BatSten1, whose seed length was comparable with that of IAC-Tatu, a Valencia peanut type (Fig. 3).

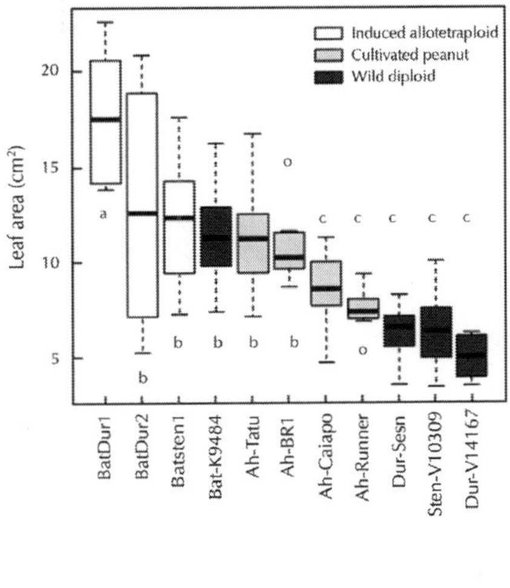

Figure 2. Box plot diagram showing the area of the first expanded leaf of wild diploids, cultivated peanut and induced allotetraploids. Boxes with same letters do not differ significantly at $P<0{\cdot}05$. The box contains 50 % of the data points. Bars across boxes represent the median. The top and bottom ends of the 'whiskers' represent the highest and lowest values observed. Circles represent outliers.

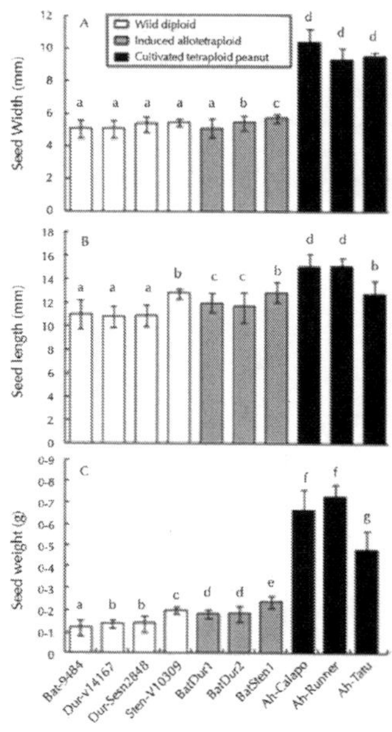

Figure 3. Seed dimensions: (A) width, (B) length and (C) weight of wild diploids, induced allotetraploids and cultivated tetraploid peanut (as indicated in the key). Bars with the same letters do not differ significantly at $P<0{\cdot}05$.

Hybridization of A. Hypogaea with the Induced Allotetraploids

Crosses were performed between five *A. hypogaea* cultivars belonging to two subspecies, *hypogaea* and *fastigiata*, and the induced allotetraploids (Table 1). Allotetraploids were used as the male parents. Pollinations carried out in the initial 18 d after the beginning of flowering generated high numbers of pegs and seeds, and these decreased considerably after that. Of a total of 3319 flowers pollinated, 234 pegs were produced, from which 133 seeds were obtained (Table 4). Peg production per pollinated flower ranged between 0·033 and 0·112. Seed production per pollinated flower ranged from 0·016 to 0·083 and showed no correlation with peanut variety or subspecies used. However, BatSten1 was the most crossable induced allotetraploid: it crossed with all cultivars of *A. hypogaea*, and produced the highest rates of pegs per pollinated flower and of seed production per pollinated flower (average of 0·0978; Table 4).

Table 4. Crossability of hybrids between peanut cultivars and induced allotetraploids

Female parent		Male parent	No. of pollinated flowers	No. of pegs	No. of seeds	No. of pegs/pollinated flower	No. of seeds/pollinated flower	No. (%) of fertile F_1 plants	Average no. of seeds/fertile F_1 plant
Cultivar	**Subspecies**								
IAC-Caiapó	*hypogaa*	BatDur1	292	24	8	0·082	0·027	6 (85·7)	5·2
		BatDur2	255	17	8	0·067	0·031	12 (92·3)	9·8
		BatSten1	166	18	10	0·108	0·060	13 (56·5)	6·0
Runner IAC 886	*hypogaea*	BatDur1	245	8	4	0·033	0·016	4 (100·0)	27·3
		BatDur2	264	13	9	0·049	0·034	11 (100·0)	12·3
		BatSten1	169	14	7	0·083	0·041	8 (61·5)	1·0
55-437 (Senegal)	*fastigiata*	BatDur1	313	16	7	0·048	0·022	12 (100·0)	3·8
		BatDur2	277	22	16	0·079	0·058	20 (100·0)	1·2
		BatSten1	217	18	13	0·083	0·060	1 (3·3)	4·0
BRS-Havana	*fastigiata*	BatDur1	202	14	7	0·069	0·035	10 (100·0)	1·8
		BatDur2	185	14	10	0·076	0·054	10 (90·9)	3·7
		BatSten1	145	15	11	0·103	0·083	0 (0·0)	N/A
BR1	*fastigiata*	BatDur1	220	11	7	0·050	0·032	7 (100·0)	0·4
		BatDur2	208	12	7	0·058	0·034	2 (25·0)	0·0
		BatSten1	161	18	9	0·112	0·056	0 (0·0)	N/A
Total			3319	234	133				

F_1 hybrids between cultivated peanut and induced allotetraploids had variable fertility, evaluated as the production of F_2 seeds. Most hybrid combinations produced a mixture of fertile and sterile F_1 plants (Table 4). Fertility varied according to cultivated peanut × allotetraploid combination. The percentage of fertile F_1 plants from allotetraploids originating from *A.duranensis* (BatDur1 or BatDur2) was generally very high (>85 %), with the exception of *A. hypogaea* 'BR1' × BatDur2 (25 % fertile). On the other hand, the percentage of fertile F_1 plants derived from the allotetraploid originating from *A. stenosperma* (BatSten1) was lower, with cultivars of the subspecies *hypogaea* ranging from 56·5 to 61·5 %, while with the subspecies *fastigiata* ranged from 0 to 3·3 % (Table 4). Also, the subspecies of cultivated peanut had a large influence on the number of seeds produced per fertile F_1 for the three allotetraploids, being higher from crosses derived from cultivars of ssp. *hypogaea* than from those derived from ssp. *fastigiata*. Notably, during the period of data collection for this study, the F_1 hybrids initially observed as being sterile remained sterile, but, in subsequent seasons, a number of them spontaneously became fertile. It was also noted that some F_1 combinations involving ssp.*fastigiata* displayed unexpected subtle phenotypic variability.

Estimated pollen viability using acetic carmine staining of wild parents, cultivated peanut and induced allotetraploids plants ranged from 89·7 to 98·8 %, and showed little variation regardless of the accession, cultivar or allotetraploid (^{2}=2637·1, d.f.=24, P<0·001) (Fig. 4). These results suggest that estimated pollen viability is not the major contributor to differences in success of F_1 production.

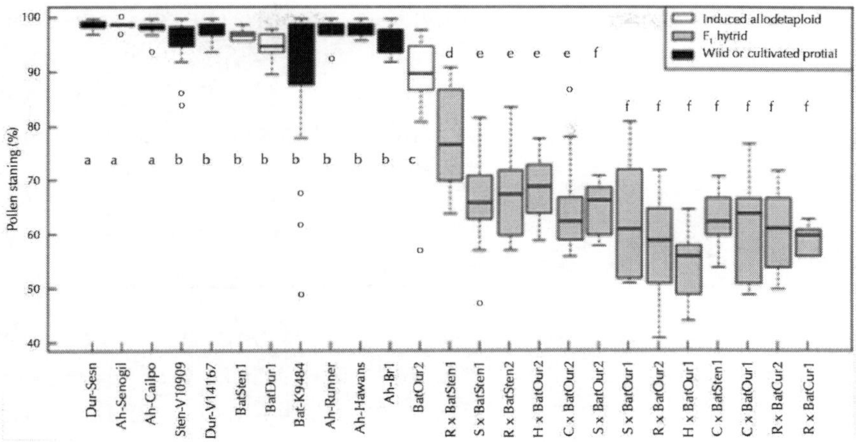

Figure. 4. Estimated pollen viability of wild accessions, peanut cultivars, induced allotetraploids and F_1hybrids between peanut cultivars and induced allotetraploids, estimated by percentage staining. On thex-axis, B=*A. hypogaea* 'BR1', C=*A. hypogaea* 'IAC-Caiapo', H=*A. hypogaea* 'BRS Havana', R=*A. hypogaea* 'IAC Runner 886', S=*A. hypogaea* '55437' (Senegal). F_1 hybrids are hybrids between cultivated peanut and induced allotetraploids. Bars with the same letters do not differ significantly at $P<0{\cdot}05$. The box contains 50 % of the data points. Bars across boxes represent the median. The top and bottom ends of the 'whiskers' represent the highest and lowest values observed. Circles represent outliers.

Pollen viability was also estimated to investigate the difference in fertility of F_1 hybrids. All hybrid plants derived from cultivated peanut × induced allotetraploids showed much lower levels of estimated pollen viability (average of 64 %) than their parents (average of 98 % for cultivated peanut and 93·3 % for allotetraploids), and this was not dependent on the peanut subspecies used as female parent or on the allotetraploid used as male parent (Fig. 4).

For the purpose of comparison with *A. hypogaea* 'Runner IAC 886' × [*A. batizocoi* K9484 × *A. duranensis* V14167]$^{4\times}$, the estimated pollen viability of a previously obtained cultivated peanut × induced allotetraploid F_1hybrid was also determined. This hybrid was the *A. hypogaea* 'Runner IAC 886' crossed with [*A. ipaensis* KG30076 × *A. duranensis* V14167]$^{4\times}$(Fávero *et al.*, 2006). This allowed the direct comparison of the influence on fertility of *A. batizocoi* compared with *A. ipaensis*, the species now considered the most probable B genome donor to cultivated peanut (Seijo*et al.*, 2004, 2007; Moretzsohn *et al.*,

2013), in hybrids involving cultivated peanut. The average pollen viability of *A. batizocoi*-derived hybrids was 77·5 %, whereas for *A. ipaënsis*-derived hybrids the average viability was 82·3 %. There was no significant difference in pollen viability between these two hybrids ($P<0{\cdot}05$).

Cytogenetic Analyses

The genomic composition of the hybrid derived from the cross between *A. hypogaea* 55437 and the allotetraploid *A. batizocoi* 'K9484' × *A. stenosperma* V10309 was analysed by GISH using total genomic DNA from*A. duranensis* V14167 (A genome, digoxigenin-labelled, green fluorescence) and *A. ipaënsis* KG30076 (B genome, biotin-labelled, red fluorescence) as probes. The expected chromosome composition was 20 chromosomes from the A genome (ten from *A. hypogaea* sub-genome A and ten from *A. stenosperma*), ten from *A. hypogaea* sub-genome B chromosomes and ten chromosomes of the K genome of *A. batizocoi* (Fig. 5A). In order to facilitate interpretation of the results, the images captured from the two fluorescent channels were overlaid (Fig. 5B). The hybrid had 40 chromosomes, of which 20 chromosomes showed green signal (from the *A. duranensis* A genome probe) and 20 showed red-orange fluorescence (from the *A. ipaënsis* B genome probe). In comparsion with the karyotype of DAPI-stained *A. hypogaea* metaphases, where 20 chromosomes of the A genome exhibit clear heterochromatic bands and 20 B genome chromosomes do not, here only ten chromosomes had no or weak centromeric DAPI bands per centromere (Fig. 5A). Comparison with the GISH image (Fig. 5B) shows that among the 20 chromosomes that have hybridized preferentially with the *A. ipaënsis* probe (red signals), ten chromosomes were DAPI banded. This indicates that these chromosomes are derived from *A. batizocoi*, which is characterized by the presence of heterochromatic DAPI bands (Seijo *et al.*, 2004). In addition, *A. batizocoi*shares a sufficient amount of repetitive DNA with the B genome to allow preferential hybridization with the *A. ipaënsis* probe. A comparison of red and orange chromosomes in close proximity (see the circle in Fig. 5B) shows that the intensity of signals is slightly weaker on *A. batizocoi* than on *A. ipaënsis* chromosomes. This observation corroborates results described earlier by Seijo and co-workers (2007), who used *A. batizocoi*and *A. ipaënsis* probes for hybridization of *A.*

hypogaea metaphases. Based on the GISH hybridization, coloured arrows were added to the image of the DAPI-stained metaphase (Fig. 5A), assigning the individual chromosomes to the species that contribute to the complex hybrid (green arrows, *A. duranensis* and *A. stenosperma*; red arrows, *A. ipaënsis*; orange arrows, *A. batizocoi*).

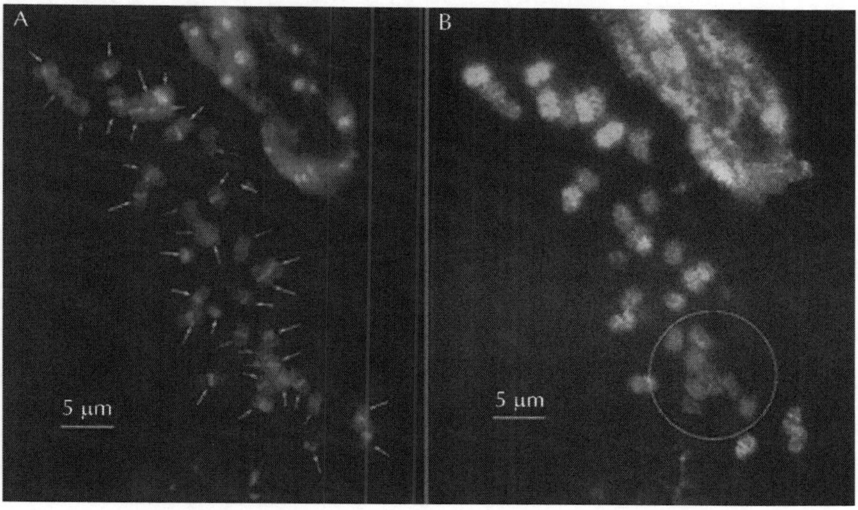

Figure. 5. Cytogenetic analysis of the hybrid between *A. hypogaea*'55437' and (*A. batizocoi*K9484 × *A. stenosperna*V10309) using GISH. Metaphase spreads of the hybrid were probed with a mixture of dig-labelled total genomic DNA from *A. duranensis* V14167 and biotin-labelled total genomic DNA from*A. ipaënsis*. (A) DAPI counterstain showing 40 chromosomes, 30 of which had strong centromeric DAPI bands, typical for *Arachis*A genome and *A. batizocoi* chromosomes. The coloured arrows indicate the nature of the individual chromosomes as derived from the combination of DAPI banding and GISH results (see text): green arrows, *A. duranensis* or *A. stenosperma*; red arrows,*A. ipaënsis*; orange arrows, *A. batizocoi*. (B) Superimposed image made from the individual FITC and Cy3 channels after GISH showing 20 green chromosomes that hybridized with the *A. duranensis* A genome probe and 20 red-orange chromosomes from hybridization with the *A. ipaensis* B genome probe. The circle comprises nine red-orange chromosomes, comprised of four *A. ipaënsis* and five *A. batizocoi* chromosomes. Comparison of hybridization intensities shows that the *A. ipaënsis* probe hybridized slightly more weakly to the *A. batizocoi*chromosomes.

Phylogenetic Analysis of A, B and K Genomes

The 16 combined TOGs yielded an aligned matrix of 9093 characters, to which were appended 74 characters from simple indel coding. Of the total characters, 8683 were constant; 289 variable characters were parsimony-uninformative and 195 were parsimony-informative. Furthermore, the matrix maximum parsimony homoplasy index was low [0·0679; PAUP* version 4.0b10 (Swofford, 2003)]. In the Bayesian analysis, the distribution of tree log likelihoods of both MrBayes runs was normal, with means of −15 706·3 and −15 705·9, respectively.

The two K genome *A. batizocoi* accessions included in the analysis clustered together, and were more closely related to *A. magna* and then toA. *ipaënsis*, both B genome species, than to the A genome species, *A. stenosperma* and *A. duranensis* (Fig. 6). With the exception of one branch within the three *A. duranensis* accessions, all branches within the Bayesian consensus phylogram possessed posterior probabilities of 1·0.

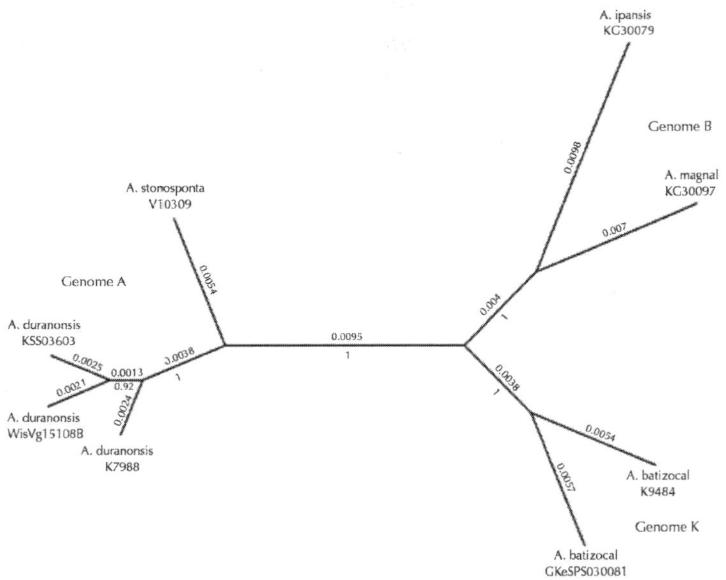

Figure. 6. Unrooted Bayesian majority rule tree inferred from 16 combined TOGs with simple indel coding. Branch lengths are given above the branches and Bayesian posterior probabilities are given below. For each genotype, the species name and accession number are given.

Testing for Ancient Hybridization Events in Arachis Species Evolution

To test the possibility that *A. batizocoi* is derived from an ancient hybridization event between A and B lineages, and reticulate evolution, some statistical analyses were performed. The phi test for evidence of recombination among the included diploid *Arachis* species was not significant ($P=0{\cdot}1408$; 114 informative sites; mean=0·126; variance=3·354E-4; observed=0·106) for the matrix comprising the concatenation of 16 TOGs and encoded indel characters from *A. batizocoi*(GKBSPSc30081), *A. duranensis* (K7988), *A. ipaënsis*, *A. magna* and *A. stenosperma*. In contrast, when *A. hypogaea* was included in the data matrix, the phi test was highly significant ($P=5{\cdot}273E\text{-}7$; 148 informative sites; mean=0·254; variance=4·852E-4; observed=0·147), reflecting the allotetraploid nature of this species, most probably comprising the A genome from *A. duranensis* and the B genome from *A. ipaënsis*. The parsimony splits analysis detected negligible evidence for reticulate evolution among the diploid *Arachis* species, whereas inclusion of *A. hypogaea* in the data set produced the expected phylogenetic network. The lack of evidence of recombination among the included diploid *Arachis*species was confirmed in the DualBrothers analysis, where a single topology predominated over the entire alignment.

DISCUSSION

The K Genome is closer to the B than the A Genome: Phylogenetic, GISH and Genetic Evidence

Arachis batizocoi has a long and interesting history in *Arachis*phylogenetic studies and of utilization in peanut breeding. Initially thought of as the B genome donor to *A. hypogaea* (Husted, 1936; Smartt *et al.*, 1978; Stalker *et al.*, 1991; Singh and Smartt, 1998), its genome was subsequently shown to be divergent from the B genome and was classified in a separate genome type, the 'K' genome (Robledo and Seijo, 2010). Since this study, the phylogenetic placement of the K genome relative to the A and B genomes has been a matter of debate. Using internal transcribed spacer (ITS) and plastid regions, Friend and co-workers (2010) found that *A. batizocoi* clustered with the *s.s.* B genome species. Similar results were obtained by Bechara and co-workers (2010), using ITS and 5.8S rDNA sequences. The analysis of Grabiele and co-workers (2012)using chloroplast DNA (cpDNA) also placed it with the B genome *s.s.*species, but basally diverging. In contrast, fluorescent *in situ* hybridization (FISH) mapping of rDNA loci has shown that K genome species are more closely related to A genome species (Robledo and Seijo, 2010).

To resolve these conflicting results, we used only a few representatives from each genome type, but, concentrating on these, we were able to use a larger data set. This consisted of 16 different intron sequences, all from different genes spread over diverse genome locations. The analysis of this data set yielded a reconstructed phylogeny of very high confidence where all the branches that define the genome types have posterior probabilities of 1·0 (Fig, 6). The tree shows that while *A. batizocoi* is distinct, it is definitively more closely related to the B than to the A genome species. Furthermore, this conclusion was supported by results from GISH. The use of *A. ipaensis* and *A. duranensis* genomic DNAs as probes on the chromosomes of a three-way intragenomic hybrid *A. hypogaea* × [*A. batizocoi* × *A. stenosperma*]^{4x} allowed the direct comparison of signals from the different component genomes on the same chromosomes set. This clearly indicated a greater affinity of the *A. batizocoi* genome for *A. ipaensis* (B genome) than for *A. duranensis* (A genome), implying a greater overall similarity of repetitive DNA content of B and K genomes. This is also consistent with genetic data that show that the genome of *A. batizocoi* recombines with the B genome of cultivated peanut in a remarkably normal way. Finally, to test the possibility that *A. batizocoi* is derived from an ancient hybridization event between A and B lineages, we tested for reticulate evolution and recombination. Using three tests, we found negligible evidence for this. We consider that these multiple lines of evidence present a significant result that effectively resolves the uncertainties of the principle phylogenetic affinities of *A. batizocoi*.

The K Genome has high Crossability, and Derived Induced Allotetraploids show the Gigas Effect

The *A. batizocoi* accession used here showed high crossability with the A genome accessions. In our experience, using the same A genome accessions, it was substantially easier to obtain primary diploid hybrids using *A. batizocoi* than *A. ipaensis* (results not shown). Highly infertile diploid hybrids were produced with all seven diploid K×A crossing combinations performed here. All individuals of two hybrid combinations exhibited symptoms typical of hybrid incompatibility, and one spontaneously died. The other six were treated with colchicine. In spite of similar efforts made to induce polyploids with these six, only three produced allotetraploids. In all of these three combinations, multiple independent allopolyploidy events were obtained. These results indicate that certain genome combinations are more crossable, certain combinations are more amenable to producing hybrid plants with normal vegetative growth, and certain combinations are more pre-disposed to induction of polyploidy. This is consistent with the barriers between species occurring at different levels (Stace, 1989). The induced allotetraploids produced were all phenotypically uniform, vigorous and highly fertile. Only immediately after germination did they show developmental irregularities. Some primary leaves of most seedlings were bleached and morphologically altered, but these abnormalities disappeared after a few weeks. The appearance of these morphological abnormalities has continued in the seedling phase for several generations, but with an overall tendency to reduce. The induced allotetraploids all showed the characteristic gigas effect (Acquaah, 2012). Being 'doubled interpecific hybrids', they have increased chromosome complements leading to gigantism, especially in leaf and overall plant size. They also show a permanent fixation of heterozygosity and hybrid vigour. Therefore, while they combine the genomes of the two parental wild species, they also display new phenotypes because of their ploidy (Leal-Bertioli *et al.*, 2012).

Compatibility of the A Genomes 'rescues' the Cultivated Peanut × Induced Allotetraploid Hybrids from the Incompatibility of the B and K Genomes

Arachis duranensis is the most likely ancestor of the A genome of *A. hypogaea*, and has good genetic compatibility with it when used as one of the parents in allotetraploids (Seijo *et al.*, 2007; Fonceka *et al.*, 2012).*Arachis stenosperma* produces hybrids with *A. duranensis* that, although reduced in fertility, were successfully used to produce a set of recombinant inbred lines (Moretzsohn *et al.*, 2005; Leal-Bertioli *et al.*, 2009; Shirasawa *et al.*, 2013). In contrast, the same accession of *A. batizocoi* used here produced highly infertile hybrids when crossed with *A. ipaensis* (Krapovickas and Gregory, 1994), the wild diploid known to be very similar to the B genome of *A. hypogaea*. Also hybrids between cultivated peanut and the amphidiploid TxAG-6 (which have *A. batizocoi*as one of the parents; Simpson *et al.*, 1993) have low fertility, although this was never precisely quantified (C.H. Simpson, Texas A&M, USA, pers. comm.).

Therefore, we anticipated that hybrids between *A. hypogaea* and the induced allotetraploids would combine generally compatible interactions on the A genomes, and generally incompatible interactions between the B and K genomes. Correspondingly, although many F_1 hybrids between *A. hypogaea* and the induced allotetraploids showed fertility, all had lower estimated pollen viability and lower fertility levels than any of the wild or cultivated accessions, or indeed the induced allotetraploids themselves. The F_1 hybrids derived from induced allotetraploids originating from *A. duranensis* showed a distinct tendency of being more fertile than F_1hybrids derived from the induced allotetraploids originating from *A. stenosperma*. Remarkably, this difference was much greater in hybrids involving *A. hypogaea* ssp. *fastigiata* than in those involving ssp. *hypogaea*cultivars. This indicates a notable genetic divergence between the two subspecies of cultivated peanut, a divergence that seems incongruent with the low levels of DNA polymorphism found between them, and the very recent species origin (Bertioli *et al.*, 2011).

Generally, it was evident that the fertility of F_1 hybrids between *A. hypogaea* and the induced allotetraploids was more influenced by the compatible genome interactions than the incompatible interactions. In other words, the compatibility of the A genome species in the *A. hypogaea* × induced allotetraploid F_1 hybrids 'rescued' them from the consequences of the incompatible B–K interactions. This was particularly emphasized by observation that the estimated pollen viabilities of *A. hypogaea* 'Runner IAC 886' × [*A. batizocoi* K9484 × *A. duranensis* V14167]$^{4\times}$ and *A. hypogaea* 'Runner IAC 886' × [*A. ipaensis* KG30076 × *A. duranensis* V14167]$^{4\times}$ were not significantly different. This observation suggests that induced allotetraploids made for peanut breeding should use at least one species that is highly related to one of the component sub-genomes of cultivated peanut. The use of two divergent species will probably produce fertile induced allotetraploids, but these will form more sterile hybrids with *A. hypogaea*.

The Variability in Fertility between Different F_1 Hybrids from the same Parents is Difficult to Explain

The variability in fertility observed between different F_1 hybrid individuals resulting from the exact same combination of *A. hypogaea* and induced allotetraploid genomes is intriguing. In all cases, the exact same combination produced some fertile and some infertile hybrids. *Arachis hypogaea*, being highly autogamous, is highly homozygous, and the induced allotetraploids should be completely so. Therefore, we would expect their F_1 progeny to be very uniform, but they definitively were not so. Intriguingly, some of the infertile F_1 individuals regained fertility in subsequent seasons. We believe that plausible explanations for these observations may lie in stochastic events following hybridization possibly involving activation of transposons, DNA methylation or other genome instability.

Concluding Remarks

In conclusion we show here that the K genome of *A. batizocoi* is distinct from the A and B *Arachis* genomes, but that it is more closely related to the B genome. The combination of high crossability, divergent genome and, when incorporated in an induced allotetraploid, fertility when crossed with cultivated peanut makes *A. batizocoi* in itself an attractive source of new alleles for improving the peanut crop, especially in crosses involving ssp. *hypogaea*. However, when the main interest is introgression of wild A genome alleles, incorporation of the A genome species in an allotetraploid with a classic B genome species will probably lead to more fertile progeny.

ACKNOWLEDGEMENTS

We thank José F. M. Valls for germplasm from the Active Germplasm Bank of Embrapa and useful discussions, Leonardo Fonseca for technical assistance, and Leandro Mesquita for assistance with greenhouse procedures. This work was supported by the Fundação de Apoio a Pesquisa do DF (FAP-DF); the Generation Challenge Programme Tropical Legumes 1; National Council for Scientific and Technological Development of Brazil (CNPq); the US National Science Foundation grant DBI 0605251 to D.R.C.; and host institutions.

REFERENCES

1. Acquaah G. 2012. Principles of plant genetics and breeding. Wiley-Blackwell.
2. Bechara M, Moretzsohn MC, Palmieri D, et al. 2010. Phylogenetic relationships in genus Arachis based on ITS and 5.8S rDNA sequences. BMC Plant Biology 10: 255.
3. Bertioli DJ, Seijo G, Freitas FO, Valls JF, Leal-Bertioli SC, Moretzsohn MC. 2011. An overview of peanut and its wild relatives. Plant Genetic Resources 9: 134–149.
4. Bruen TC, Philippe H, Bryant D. 2006. A simple and robust statistical test for detecting the presence of recombination. Genetics 172: 2665–2681.
5. Burow M, Simpson C, Starr J, Paterson A. 2001. Transmission genetics of chromatin from a synthetic amphidiploid to cultivated peanut (Arachis hypogaea L.). Broadening the gene pool of a monophyletic polyploid species. Genetics 159: 823–837.

6. Choi H-K, Luckow M, Doyle J, Cook D. 2006. Development of nuclear genederived molecular markers linked to legume genetic maps. Molecular Genetics and Genomics 276: 56–70.
7. Creste S, Yulmann Neto A, Figueira A. 2001. Detection of single sequence repeat polymorphisms in denaturing polyacrylamide sequencing gels by silver staining. Plant Molecular Biology Reporter 19: 299–306.
8. Drummond AJ, Ashton B, Buxton S, et al. 2011. A: Geneious v5.4. Available from http://www.geneious.com/. Edgar R. 2004. MUSCLE: multiple sequence alignment with high accuracy and high throughput. Nucleic Acids Research 32: 1792–1797.
9. Ewing B, Green P. 1998. Base-calling of automated sequencer traces using Phred II. Error probabilities. Genome Research 8: 186–194. Fa´vero AP, Simpson CE, Valls FMJ, Velo NA. 2006. Study of evolution of cultivated peanut trough crossability studies among Arachis ipae¨nsis, A duranensis and A hypogaea. Crop Science 46: 1546–1552.
10. Fonceka D, Tossim H, Rivallan R, et al. 2012. Construction of chromosome segment substitution lines in peanut (Arachis hypogaea L.) using a wild synthetic and QTL mapping for plant morphology. PLoS One 7: e48642.
11. Ferreira ME, Grattapaglia D. 1998. Introduc¸a˜o ao uso de marcadores moleculares em ana´lise gene´tica, 3rd edn. Brası´lia: Embrapa-Cenargen, Friend SA, Quandt D, Tallury SP, Stalker HT, Hilu KW. 2010. Species, genomes, and section relationships in the genus Arachis (Fabaceae): a molecular phylogeny. Plant Systematics and Evolution 290: 185–199.
12. Grabiele M, Chalup L, Robledo G, Seijo G. 2012. Genetic and geographic origin of domesticated peanut as evidenced by 5S rDNA and chloroplast DNA sequences. Plant Systematics and Evolution 2012: 1151–1165.
13. Halward TM, Wynne JC, Stalker HT. 1991. Recurrent selection progress in a population derived from an interspecific peanut cross. Euphytica 52: 79–84.
14. Holbrook CC, Timper P, Culbreath AK, Kvien CK. 2008. Registration of 'Tifguard' peanut. Journal of Plant Registrations 2: 2.

15. Huelsenbeck JP, Larget B, Alfaro ME. 2004. Bayesian phylogenetic model selection using reversible jump Markov Chain Monte Carlo. Molecular Biology and Evolution 21: 1123–1133.
16. Huson DH, Bryant D. 2006. Application of phylogenetic networks in evolutionary studies. Molecular Biology and Evolution 23: 254–267.
17. Husted L. 1936. Cytological studies on the peanut. Arachis. II – Chromosome number, morphology and behavior, and their aplication to the problem of the origin of the cultivated forms. Cytologia 7: 396–423.
18. Kochert G, Stalker HT, Gimenes M, Galgaro L, Lopes CR, Moore K. 1996. RFLP and cytogenetic evidence on the origin and evolution of allotetraploid domesticated peanut, Arachis hypogaea (Leguminosae). American Journal of Botany 83: 1282–1291.
19. Krapovickas A, Gregory W. 1994. Taxonomia del genero Arachis (Leguminosae). Bonplandia 8: 1–186.
20. Leal-Bertioli SCM, Jose´ ACFV, Alves-Freitas DMT, et al. 2009. Identification of candidate genome regions controlling disease resistance in Arachis. BMC Plant Biology 9: 112.
21. Leal-Bertioli SCM, Bertioli DJ, Guimara˜es PM, et al. 2012. The effect of tetraploidization of wild Arachis on leaf morphology and other drought-related traits. Environmental and Experimental Botany 84: 17–24.
22. Linsley EG, Cazier MA. 1963. Further observations on bees which take pollen from plants of the genus Solanum. Pan Pacific Entomology 39: 1–18.
23. Minin VN, Dorman KS, Fang F, Suchard MA. 2005. Dual multiple changepoint model leads to more accurate recombination detection. Bioinformatics 21: 3034–3042.
24. Moretzsohn MC, Hopkins MS, Mitchell SE, Kresovich S, Valls JFM, Ferreira ME. 2004. Genetic diversity of peanut (Arachis hypogaea L.) and its wild relatives based on the analysis of hypervariable regions of the genome. BMC Plant Biology 4: 11.
25. Moretzsohn MC, Leoi L, Proite K, et al. 2005. A microsatellite-based, generich linkage map for the AA genome of Arachis (Fabaceae). Theoretical and Applied Genetics 111: 1060–1071.

26. Moretzsohn MC, Gouvea EG, Inglis PW, Leal-Bertioli SCM, Valls JFM, Bertioli DJ. 2013. A study of the relationships of cultivated peanut (Arachis hypogaea) and its most closely related wild species using intron sequences and microsatellite markers. Annals of Botany 111: 113–126.
27. Nagy E, Chu Y, Guo Y, Khanal S, et al. 2010. Recombination is suppressed in an alien introgression in peanut harboring Rma, a dominant root-knot nematode resistance gene. Molecular Breeding 26: 357–370.
28. Pijnacker LP, Ferwerda MA. 1984. Giemsa C-banding of potato chromosomes. Canadian Journal of Genetetics and Cytology 26: 415–419.
29. Rambaut A, Suchard MA, Xie D, Drummond AJ. 2014. Tracer v1.6. Available from http://beast.bio.ed.ac.uk/Tracer.
30. Robledo G, Seijo G. 2010. Species relationships among the wild B genome of Arachis species (section Arachis) based on FISH mapping of rDNA loci and heterochromatin detection: a new proposal for genome arrangement. Theoretical and Applied Genetics 121: 1033–1046.
31. Robledo G, Lavia G, Seijo G. 2009. Species relations among wild Arachis species with the A genome as revealed by FISH mapping of rDNA loci and heterochromatin detection. Theoretical and Applied Genetics 118: 1295–1307.
32. Ronquist F, Teslenko M, van der Mark P, et al. 2012. MrBayes 3.2: efficient Bayesian phylogenetic inference and model choice across a large model space. Systematic Biology 61: 539–542.
33. Schwarzacher T, Heslop-Harrison JS. 2000. Practical in situ hybridization. Oxford: BIOS Scientific Publishers. Seijo JG, Lavia GI, Fernandez A, Krapovickas A, Ducasse D, Moscone EA. 2004. Physical mapping of the 5S and 18S-25S rRNA genes by FISH as evidence that Arachis duranensis and A. ipae¨nsis are the wild diploid progenitors of A. hypogaea (Leguminosae). American Journal of Botany 91: 1294–1303.
34. Seijo G, Lavia GI, Fernandez A, et al. 2007. Genomic relationships between the cultivated peanut (Arachis hypogaea, Leguminosae) and its close relatives revealed by double GISH. American Journal of Botany 94: 1963–1971.

35. Simmons MP, Ochoterena H. 2000. Gaps as characters in sequence-based phylogenetic analyses. Systematic Biology 49: 369–381.
36. Shirasawa K, Bertioli DJ, Varshney RK, et al. 2013. An integrated consensus map of cultivated peanut and wild relatives revealed structures of the A and B genomes of Arachis and divergences with other of the legume genomes. DNA Research 20: 173–184.
37. Simpson CE. 1991. Pathways for introgression of pest resistance into Arachis hypogaea L. Peanut Science 18: 22–26.
38. Simpson CE, Starr JL. 2001. Registration of 'COAN' peanut. Crop Science 41: 918.
39. Simpson CE, Nelson SC, Starr LJ, Woodard KE, Smith OD. 1993. Registration of TxAG-6 and TxAG-7 peanut germplasm lines. Crop Science 33: 1418.
40. Simpson CE, Starr JL, Church GT, Burow MD, Paterson AH. 2003. Registration of 'NemaTAM' peanut. Crop Science 43: 1561.
41. Singh AK, Smartt J. 1998. The genome donors of the groundnut/peanut (Arachis hypogaea L.) revisited. Genetic Resources and Crop Evolution 45: 113–118.
42. Smartt J, Gregory W, Gregory M. 1978. The genomes of Arachis hypogaea. 1. Cytogenetic studies of putative genome donors. Euphytica 27: 665–675.
43. Stace CA. 1989. Plant taxonomy and biosystematics. New York: Chapman & Hall.
44. Staden R. 1996. The Staden sequence analysis package. Molecular Biotechnology 5: 233–241.
45. Stalker HT, Dhesi JS, OParry DC. 1991. An analysis of the B genome species Arachis batizocoi (Fabaceae). Plant Systematics and Evolution 78: 238–246.
46. Swofford DL. 2003. PAUP*: phylogenetic analysis using parsimony (*and other methods), 4.04 Beta. Sunderland, MA: Sinauer Associates.
47. Tallury SP, Hilu KW, Milla SR, et al. 2005. Genomic affinities in Arachis section Arachis (Fabaceae): molecular and cytogenetic evidence. Theoretical and Appled Genetics 111: 1229–1237.
48. Vaidya G, Lohman DJ, Meier R. 2011. SequenceMatrix: concatenation software for the fast assembly of multi-gene

datasets with character set and codon information. Cladistics 27: 171–180.

49. Vale FXR, Fernandes Filho E, Liberato JR. 2001. Quant – Quantificac¸a˜o de doenc¸as: versa˜o 1.0.1., p. 1.0.1. Vic¸osa: UFV.
50. Valls JFM, Costa LC, Custodio AR. 2013. A novel trifoliolate species of Arachis (Fabaceae) and further comments on the taxonomic section Trierectoides. Bonplandia 22: 91–97.
51. Valls JFM, Simpson CE. 2005. New species of Arachis L. (Leguminosae) from sBrazil, Paraguay and Bolivia. Bonplandia 14: 35–63.

Chapter 6

Peanut (Arachis Hypogaea) Expressed Sequence Tag Project: Progress and Application

Suping Feng[1,2] Xingjun Wang,[3] Xinyou Zhang,[4] Phat M. Dang,[5] C. Corley Holbrook,[6] Albert K. Culbreath,[1] Yaoting Wu,[2] and Baozhu Guo[1,7]

[1]Department of Plant Pathology, University of Georgia, Tifton, GA, USA

[2]College of Bioscience and Biotechnology, Qiongzhou University, Sanya, China

[3]High-Tech Research Center, Shandong Academy of Agricultural Sciences, Jinan, China

[4]Henan Academy of Agricultural Sciences, Zhengzhou, China

[5]National Peanut Research Laboratory, USDA-ARS, Dawson, GA, USA

[6]Crop Genetics and Breeding Research Unit, USDA-ARS, Tifton, GA, USA

[7]Crop Protection and Management Research Unit, USDA-ARS, Tifton, GA, USA

ABSTRACT

Many plant ESTs have been sequenced as an alternative to whole genome sequences, including peanut because of the genome size and complexity. The US peanut research community had the historic 2004 Atlanta Genomics Workshop and named the EST project as a main priority. As of August 2011, the peanut research community had deposited 252,832 ESTs in the public NCBI EST database, and this resource has been providing the community valuable tools and core foundations for various genome-scale experiments before the whole genome sequencing project. These EST resources have been used for marker development, gene cloning, microarray gene expression and genetic map construction. Certainly, the peanut EST sequence resources have been shown to have a wide range of applications and accomplished its essential role at the time of need. Then the EST project contributes to the second historic event, the Peanut Genome Project 2010 Inaugural Meeting also held in Atlanta where it was decided to sequence the entire peanut genome. After the completion of peanut whole genome sequencing, ESTs or transcriptome will continue to play an important role to fill in knowledge gaps, to identify particular genes and to explore gene function.

INTRODUCTION

Legumes are a diverse and important family of angiosperms. With more than 650 genera and 18,000 species, legumes are the third largest family of higher plants and are second only to grasses in agriculture [1]. Peanut is one of the major economically important legumes and is widely grown in the southern United States, and countries like China, India, and in South and Central America and Africa. On a global basis, peanut is a major source of protein and vegetable oil for human nutrition. China and India together produce almost two-thirds of the world's peanuts, and the US produces about 6% (http://faostat.fao.org/). Nearly two-thirds of global production is crushed for oil, and the remaining one third is consumed as food. From 2004 to 2006, peanuts were grown on an average of 21.49 million hectares worldwide with production totaling 32.98 million metric tons (http://www.nass.usda.gov/Publications/Ag_Statistics/2008/Chap03.pdf). During the same time, the US peanut averaged 571 thousand hectares with production of 1.9 million metric tons. Peanut production also has a significant role in sustainable agriculture in terms of global food security and nutrition, fuel and energy, sustainable fertilization, and enhanced agricultural productivity as a rotation crop. Peanut offers numerous health benefits, but is also one of the primary food allergens. High production cost is another challenge to peanut farmers as exemplified by chemical control of foliar diseases. Food safety concern due to aflatoxin contamination of peanut is another important issue threatening human health.

Genomic research can provide new tools and resources to enhance crop genetic improvement and production [2]. However, genomic research in peanut is far behind those in other crops, such as maize, soybean, wheat, sorghum, and potato due to the shortage of essential genomic infrastructure, tools, and resources. Although peanut is an economically and nutritionally important crop globally, it was virtually unexplored at the genome level. On March 22-23, 2004, scientists with expert knowledge of critical fields in genetics and plant molecular biology participated in a workshop hosted by the Peanut Foundation/American Peanut Council in Atlanta, GA, USA. These scientists reviewed the status of peanut genomic research, which was documented in the book entitled Legume Crop Genomics [3] under the auspices of the US Legume Crop Genome Initiative (LCGI). The Peanut Genome Initiative (PGI) was launched at this workshop. An advisory committee, representing the broad interests of industry and the peanut research community, was selected to guide the growth of the PGI. A Strategic Plan for the Peanut Genome Initiative 2004–2008 was developed that outlined research goals, objectives, performance measures, and significant near-term milestones in the advancement of this emerging science (http://www.peanutbioscience.com/). Development of expressed sequence tag (EST) database was on the top of the priorities.

There have been tremendous developments and applications from the EST concept proposed by Adams et al. [4]; however, only limited but encouraging progress has been made in peanut. Among the plant genomes, ESTs are currently the most widely sequenced nucleotides in terms of the number of sequences and the total nucleotide count. ESTs provide a robust sequence resource that can be exploited for gene discovery, genome annotation, and comparative genomics. In this paper, we will summarize the progress that has been made and the applications that have been developed, particularly the chronological events of development of Peanut Genome Initiative (PGI) from EST to whole genome sequencing as archived inhttp://www.peanutbioscience.com/.

THE PAST AND PRESENT OF PEANUT EST PROJECTS

EST was an important research area of functional genomics during the early days when only a few plant genomes were sequenced and before the development of the second generation of high throughput sequencing technology. It was especially important for major crops or economically significant plants with large genomes, such as peanut. EST sequencing was a feasible approach for researcher to gain sequence information for gene cloning, gene expression analysis, and molecular marker development. However, at the dawn of launching the PGI at the historic Atlanta Workshop in March 2004, there was virtually no peanut EST in public databases like NCBI dbEST (GenBank) (http://www.ncbi.nlm.nih.gov/dbEST/dbEST_summary.html/). For example, for plant ESTs in NCBI dbEST (February 15, 2003), wheat (Titicum aestivum) had the most with 415,589, barley (Hordeum vulgare) was the second with 314,884, soybean (Glycine max) had 308,564, maize (Zea mays) had 198,034, rice (Oryza sativa) had 130,775, Arabidopsis thaliana had 178,464, Sorghum bicolor had 84,712, and potato (Solanum tuberosum) had 94,423. After the launching of the PGI, Luo et al. [5] reported the first peanut EST project and released 1,345 ESTs to GenBank EST database in May 2003 (accession numbers CD037499 to CD038843).

Peanut EST projects started very slow. Two years after the PGI was launched (January 20, 2006), peanut had 2,171 ESTs in the GenBank. At the same time, there were 7,596,977 human ESTs. Among plants, maize had the most ESTs with 662,884,and wheat was second with 600,205. Arabidopsis thaliana had 421,027, rice had 407,545, soybean had 356,808, and even Medicago truncatula had 221,123. Peanut was dramatically low on the list for such an important crop.

It has been a long journey for the peanut research community to reach a comparable number that we have today (March 1, 2012). Including wild diploid peanut, the peanut community has deposited

252,832 ESTs to NCBI EST database. There are 178,490 for A. hypogaea (including 745 for subsp. fastigiata), 35,291 for A. duranensis, 32,787 for A. ipaensis, and 6,264 for A. stenosperma. However, there are a total of 72,098,708 ESTs in the public database, and 8,315,295 are human ESTs. Maize has 2,019,114, soybean has 1,461,624, rice has 1,252,989, wheat has 1,073,877, barley has 501,838, upland cotton (Gossypium hirsutum) has 297,239, and potato has 249,761. Before the completion of peanut whole genome sequencing, sequencing large numbers of ESTs can create a formidable resource for studies in both biodiversity and gene discovery. Sequence analysis tools have extended the scope of EST analysis into the fields of proteomics, marker development, and genome annotation. Although EST collections certainly are not intended to substitute for a whole genome sequences, the EST resource forms the core foundations for various genome wide experiments, particularly for microarray gene expression study, marker development, and genetic map construction, which could also be useful to assist research to assemble the whole genome.

Peanut ESTs were derived from different tissues, different growth stages and under different abiotic and biotic stresses [5–12]. For instance, Luo et al. [5] construct two cDNA libraries from leaves of the peanut cultivar Tifrunner (C34-24) [13] and A13. This was the first report of ESTs in cultivated peanut. Proite et al. [6] obtained 6,264 high-quality ESTs from four cDNA library leaves and roots of A. stenosperma and developed 188 SSR markers. Alves et al. [14] selected ten EST sequences with homologies of interest and obtained 8 SNP markers based on the single-base primer extension method. Recently, Koilkonda et al. [15] developed 3,187 EST-SSR markers, which were obtained from 10,102 potential nonredundant EST sequences, including 3,445 contigs and 6,657 singletons, generated from cDNA libraries of the gynophores, roots, leaves, and seedlings of cultivated peanut.

Yan et al. [16] constructed a cDNA library of mid-maturation stage cotyledons of peanut cultivar Shanyou 523 and obtained 414 EST sequences. Seventeen percent (17%) of these ESTs-encoded seed storage proteins, including 5 arachin, 2 conarachin, and 6 conglutin-like proteins encoding sequences [17]. Four new members of arachin family and two new members of conglutin family were discovered in peanut.

A full-length cDNA library from immature peanut seed was reported by Wang et al. [18]; over 17,000 ESTs were sequenced, and 10,000 ESTs were deposited to the NCBI EST database [10]. From these sequences, large numbers were storage protein-encoding genes, genes encoding transcription factors, and stress- or pathogen-related proteins. Proteins involved in protein destination and storage represented the largest group and accounted for 28.8% of the total proteins. Genes involved in metabolism accounted for 12.5% of the total ESTs. A large proportion of ESTs encoded proteins with unknown function (20.1%) or proteins with no hits (13.9%) in the data base providing the opportunity for novel gene discovery [10].

To gain a better understanding of the high oleic acid accumulation in a high oleic acid peanut variety E12, a cDNA library was constructed and obtained 12,501 ESTs. A total of 4,074 unigenes were generated from these ESTs. Proteins encoded by these ESTs are involved in a variety of biological processes including aminoacid and carbohydrate metabolism, energy metabolism, transcription, protein translation, and transportation [19, 20]. These EST sequences provide valuable information for gene cloning, such as genes encoding key enzymes for fatty acid and seed storage protein biosynthesis.

In order to understand bacterial wilt resistance in peanut, cDNA libraries were constructed from normal leaves and leaves challenged with pathogens using the peanut line 06-4104 with high oil content (56%) and resistance to bacterial wilt. The differentially expressed genes between normal and pathogen-challenged plants were identified. From the EST sequence information (over 25,000 ESTs), 5,920 unigenes from the normal leaf library and 7,507 unigenes from the pathogen challenged library were identified [21]. Xiao et al. [22] reported construction of 5 new cDNA libraries using a novel line with high oil content and high resistance to bacterial wilt disease. EST sequencing was carried out using these libraries and a total of 63,207 ESTs, representing 14,547 uni-ESTs, were generated. Recently, a total of 63,234 ESTs were deposited to the public database (GenBank dbEST) with accession numbers of JK146921 to JK210154.

GENE CLONING BASED ON EST SEQUENCES

One important aim for EST sequencing is to clone the full-length sequences of genes with agronomic value. Fatty acid synthesis is one of the most important traits for the oil crop peanut. The first enzyme complex in fatty acid synthesis in plants is ACCase which is composed of 4 subunits, BCCP, BC, -CT, and -CT. With the help of peanut seed EST sequence information combined with homology cloning, the full-length ORF of these four subunits was cloned from cultivated peanut. The multifunctional ACCase, one peptide containing three functional domains, was also identified from peanut [23]. Plant fatty acid biosynthesis is catalyzed by type II fatty acid synthase (FAS) in plastids and mitochondria. Type II FAS complex contains several enzyme and an important protein, the acyl-carrier protein (ACP). Genes encoding ACP, malonyl-CoA: ACP transacylase, -ketoacyl-ACP synthase, -ketoacyl-ACP reductase, -hydroxyacyl-ACP dehydrase and enoyl-ACP reductase were isolated using a similar strategy. One to five members of genes encoding each enzyme were identified [24]. Five different types of ACP genes that showed little sequence similarity were cloned [25]. Oleosin is an important component of plant oil bodies. The expression and accumulation level of a specific oleosin could influence the morphology of oil bodies and seed oil content [26]. There are 284 ESTs that showed high sequence homology with oleosin from other plant species. These ESTs form 6 contigs and represent 6 subfamilies of peanut oleosins. The full-length ORF of these oleosin genes was also cloned (X. Wang, unpublished data).

Several peanut storage proteins are allergens. From the developed ESTs, 8 allergen genes, Ara h1 to Ara h8 were identified [27]. LEA (late embryogenesis abundant) proteins which play key roles in the protection of seed desiccation accumulate abundantly in seed during late embryo development. Su et al. [28] reported the identification of 19 LEA proteins belonging to 8 different subfamilies. The full-length ORF of many other genes involved in different biological processes, for example, the anthocyanin biosynthesis-related genes were also cloned in the EST project (X. Wang, unpublished data). Based on the EST sequence information several transcription factors including AhLEC1, AhDREB, and AhNAC were cloned and characterized [24, 29, 30]. Promoters of genes with special expression patterns, for example, seed or developmental stage-specific expression, have potential applications in gene engineering research. Cloning of these genes and their promoters is of great interest. Several seed-specific or developmental preferential genes were identified and some of their promoters were characterized ([10]; X. Wang and Han Xia, unpublished data). Based on a pericarp and testa-specific EST, the full-length ORF of a cysteine protease gene was cloned [31].

Genes that may play roles in general disease resistance or in a specific disease are a great interest to researchers. The plant defensin is a key member of the plant immune system, which could play roles in protecting plants from pathogen attack. Based on the EST sequences of defensin together with RACE amplification of the full-length ORF of peanut defensin was cloned [32]. Using the information of an A. flavus responsive EST, a full length peanut PR-10 gene was cloned. This protein showed a low level of sequence similarity with the previously reported AhPR-10 and suggested the identification of a novel peanut PR-10 [33]. From the EST sequencing project, Zhao et al. [34] reported that 237 ESTs showed homology with plant lipid transfer protein (LTP) coding genes. These ESTs could form 5 different contigs representing five types of LTP. Plant LTPs were reported to play roles in a variety of biological processes such as development and reproduction, resistance to adverse environment as well as disease resistance.

GENE EXPRESSION STUDY USING THE EST RESOURCES

Microarray analysis is a powerful tool for global gene expression profiling. The first peanut microarray study was carried out by Luo et al. [35] to investigate the differentially expressed genes in peanut in response to A. parasiticus infection and drought stress. This microarray was made of cDNA clone spot-array using ESTs from two cDNA libraries [35, 36]. Later, an oligonucleotide microarray containing more EST sequence information was also designed and used for peanut gene profiling [37]. An oligonucleotide microarray containing 15,744 unique probes was created from 49,205 peanut ESTs. A total of 36,766 probes were designed using the server-based eArray platform from Agilent Technologies, an in situ-synthesized microarray platform. A full description of the array is available at the NCBI GEO (gene expression omnibus) database with accession GPL6661. Recently, Guo et al. [38] reported the use of long oligonucleotide sequences in gene expression profiling experiments to identify candidate genes that confer resistance to Aspergillusinfection due to upexpression in response to fungal infection. The array platform description can be found at the NCBI GEO database (accession GPL13178). Briefly, oligonucleotides ranging from 60 to 70 mer were designed at the J. Craig Venter Institute (JCVI). The total number of oligonucleotides spotted on the microarray was 6,932, the ultra-GAPs glass slides with 3 replications of each oligonucleotide at different locations on the slide.

There is another report on peanut microarray study [10], which made a cDNA microarray using the EST sequences obtained from immature seeds. This microarray was used to analyze differential gene expression among peanut tissues and organs. Gene expression patterns were also analyzed during peanut seed development. Recently, Zhuang et al. [39] made an oligonucleotide gene chip using more than 100,000 unigenes derived from 454 sequences, using this gene chip for peanut genes expression analysis especially for genes that may be involved in A. flavus tolerance [39].

EST-SSR MARKER DEVELOPMENT

Guo laboratory and collaborator developed a large number of EST-SSR (simple sequence repeat) markers from peanut EST sequences [40, 41]. About 24,000 were analyzed for SSR discovery. A total number of 881 EST-SSRs were identified, and 251 of them could be successfully amplified from peanut. Most of these SSRs exhibited polymorphism in the wildtype peanut; however, there were only a small number of the SSRs showed polymorphism in cultivated peanut [40]. In addition, 740 SSRs were discovered from 20,160 new EST sequences derived from cultivated peanut pod [42], which are currently not deposited in the database. The amplification and polymorphism of some SSRs were tested both in cultivated and wildtype peanut. Using ESTs from immature peanut seed cDNA library, Song et al. [11] identified 841 EST-SSRs. Part of these SSRs were used to analyze the polymorphism among cultivated and wild type peanuts. From 63,207 ESTs derived from 5 cDNA libraries constructed using bacterial wilt resistance peanut line "06-4104", 2,643 EST-SSR loci were identified [22]. Qin et al. [41] reported the largest collection of SSRs, a total of 4,576 SSR markers from three sources: published SSR markers, newly developed SSR markers from ESTs and from bacterial artificial chromosome (BAC) end-sequences for construction of a genetic linkage map, and QTL analysis of TSWV resistance.

TRANSFERABILITY AND COMPARATIVE GENOMICS

Gene content and order are highly conserved among closely related species as revealed by comparative genetics. Sequence data obtained from several crop plants indicated homology existing between genomes of two or more closely related genera/species. EST-SSRs have the effective transferability across genera/species in many crops [43–47]. EST-SSR markers have a higher transfer rate than SSR markers from genomic sequences due to conservation of transcribed regions among related species [48, 49]. Mace et al. [50] obtained 51 SSR markers from the Leguminosae family by using in silico method and also tested 27 diverse Arachis accessions, and 18 revealed polymorphisms. Varshney et al. [51] constructed the first SSR-based cultivated peanut map by using partial SSR markers from the diploid AA genome map [52], and legume anchor markers were developed and compared to maps fromArachis, Lotus, and Medicago. Moretzsohn et al. [53] constructed a B-genome linkage map using microsatellite markers developed for other Arachis species and showed high transferability (81.7%). This B-genome map was compared to the A-genome map using 51 common markers. Foncéka et al. [54] reported that synteny analysis between A and B genomes revealed an overall good collinearity of the homeologous LGs.

CONSTRUCTION OF GENETIC LINKAGE MAP

Development of a genetic linkage map has been a core effort in the international peanut research community, but progress was slow until 2005. A detailed review was presented by Guo et al. [2]. The first application of EST-SSR markers was to construct a linkage map by using an F2 population obtained from a cross between two diploid species with AA genomes (A. duranensis and A. stenosperma), and this map contained 170 SSR loci (from both genomic SSRs and EST-SSRs) on 11 linkage groups (LGs) covering 1231 cm [52]. Alves et al. [14] located five candidate genes for resistance on this genetic map. An advanced version of this map was developed later by Leal-Bertioli et al. [55], by further saturation with 369 marker loci. A diploid B-genome map was also established from an F2 population of A. ipaensis × A. magna covering 1294.4 cm and 149 marker loci, of which 25% of the markers were developed from cDNA libraries [53].

There are very few genetic maps available based on cultivated peanuts because of low levels of polymorphism. The first SSR-based genetic linkage map for cultivated peanut was constructed by Varshney et al. [51] with 135 marker loci into 22 LGs, covering 1,270.5 cm. This map was further updated with 191 SSR loci onto 20 LGs with 1,785 cm genome coverage [56]. Foncéka et al. [54] published an SSR-based tetraploid map of the BC1F1 population of "Fleur 11" × (A. ipaensis × A. duranensis)^{4x} with 298 loci onto 21 LGs. Another SSR-based composite genetic linkage map of cultivated peanut was published, based on three RIL populations constructed from three crosses with one common female parent, contained 22 LGs with 175 SSR markers [57]. Mondal et al. [58] constructed a genetic linkage map, containing 24 LGs with 109 SSR markers, covering 882.9 cm by using a total of 164 recombinant inbred lines derived from rust resistant (VG 9514) and susceptible (TAG 24) cultivated groundnut parents. Since the turning point of the 2004 Atlanta Workshop, more EST-SSRs are available for community use, but marker density is still very low, and more markers are needed, particularly SNP markers. The most recent map by Qin et al. [41] reports an integrated genetic linkage map of cultivated peanut constructed from two RIL (recombinant inbred line) populations, derived from the cross "Tifrunner" × "GT-C20" and the cross "SunOleic 97R" × "NC94022." "Tifrunner" is a runner market-type peanut (Arachis hypogaea L. subsp. hypogaea var. hypogaea) cultivar with a high level of resistance to TSWV (tomato spotted wilt virus), and moderate resistance to early (Cercospora arachidicola) and late leaf spot (Cercosporidium personatum). "GT-C20" is a Spanish-type breeding line (A. hypogaea L. subsp. fastigiata var. fastigiata) and highly susceptible to TSWV and leaf spots but resistant to aflatoxin contamination. "SunOleic 97R" [59] is from a BC_4F_5 selection of a cross of "SunOleic 95R" [60] and "Sunrunner" [61] with high oleic fatty acid but susceptible to TSWV. The breeding line "NC94022" has been reported to have a high level of field resistance to TSWV [62], a selection from a cross between N91026E (an early maturity Virginia type) and a tan-seeded component selected from PI 576638, a hirsuta type line. A total of 4576 simple sequence repeat (SSR) markers were used for screening polymorphisms. Two CAPS (cleaved amplified polymorphic sequence) markers were also included to differentiate ahFAD2A alleles [63] and ahFAD2B alleles [64]. A total of 324 markers were anchored on this integrated map covering 1,352.1 cm with 21 linkage groups (LGs). This map defined 7 homoeologous groups, and 17 LGs were aligned to corresponding A-subgenome (A1 to A10) or B-subgenome (B1 to B4, B7, B8, and B9) of diploid progenitors, which could assist the assembly of peanut whole genome sequencing project. The Peanut Genome Consortium (PGC), an extension of the International Peanut Genome Initiative (IPGI) has initiated a project titled "Creating a Better Future through Global Food Security" and decided to sequence these two RIL populations (http://www.peanutbioscience.com/). Further two major QTLs for TSWV resistance were also identified for each RILs, which illustrated the application of this map [41].

PEANUT GENOME AND WHOLE GENOME SEQUENCING

Cultivated peanut is an allotetraploid 2n=4x=40 with little polymorphism at the molecular level [65, 66] as indicated by using traditional markers such as RAPD and RFLP [67, 68]. Pairing in A. hypogaea is generally bivalent, with occasional higher-order associations found in crosses among different market types [69]. Cultivated peanut is considered to have originated from a single recent polyploidization event [70, 71], unlike many other natural polyploidy species for which multiple polyploidization events have been identified. In cultivated peanut, the most likely wild diploid progenitors are A. duranensis (the A-genome) and A. ipaensis (the B-genome) [70]. The peanut genome (2,800 Mb/1C) is large in comparison to the plant models, Arabidopsis (128 Mb), rice (420 Mb), Medicago (500 Mb), corn (2,500 Mb), and even soybean

(1,100 Mb). A. thaliana was estimated to have 27,000 genes; rice is predicted to have 30,000 to 50,000 genes [72, 73]; soybean is about 60,000 genes [74]. While the peanut genome probably has a gene number similar to the small-genome species, the large genome size and the complexity makes it difficult to sequence and assemble the whole genome. Therefore, any large-scale discovery, isolation, and deciphering of gene functions for marker development in peanut had been done by relying on other, less cost intensive methods such as the EST project. The EST project focused solely on the development of genomic tools and resources for the peanut community to ensure continued advance in understanding the biology of this important crop species and continued improvement through crop breeding. Moreover, it also constructed a bridge between the peanut genomic research and research on other important legumes for fluid movement of information on gene location and function from other legume genomes to peanut and vice versa, and facilitated exploitation of the high conservation of gene order for comparative functional genomics.

Next-generation sequencing technology and cost reductions have enabled complete genome sequences for many plants, such as rice (2005), maize (2009), sorghum (2009), soybean (2010), and potato (2011). Because of the genome size, sequencing of the entire peanut genome was not feasible until recently. The Peanut Foundation and the American Peanut Council on behalf of international peanut research community and Peanut Genome Consortium (PGC) discussed pursuing a peanut genome sequencing project and related issues on July 12, 2010, at Clearwater, FL, USA, and made the decision to send a delegation to China to initiate discussions with Chinese peanut collaborators for a possible joint sequencing project along with other international partners. Two members from the executive committee Victor Nwosu, Plant Science Program Manager, Global Chocolate Science & Technology of Mars Chocolate, NA, and Baozhu Guo, plant pathologist, USDA-ARS Crop Protection and Management Research Unit traveled to China from September 2–12, 2010 to visit Chinese peanut genomic research collaborator. X. Zhang, Henan Academy of Agricultural Sciences and Peanut Breeder, was the local host and arranged a meeting with Chairman Fuhe Luo, and meeting with others, including X. Wang of Shandong Academy of Agricultural Sciences and Dr. Da Luo of Sun Yat-Sen University. A proposal was made to collaborate and pull resources together to sequence both the tetraploid and diploid peanut genomes. This trip led to another important Atlanta meeting, Peanut Genome Project Inaugural Meeting (http://www.peanutbioscience.com/peanutgenomeproject.html) on December 8, 2010. The delegation briefed the Executive Committee and the PGI agreed in principle to move forward with this collaborative effort to sequence peanut genomes.

The US PGI Committee discussed pursuing a peanut genome sequencing project and related issues on December 8, 2010, at the Winter Conference, Atlanta, and on January 12, 2011, at the Plant and Animal Genome Conference, San Diego, CA, USA, organized by Dr. Howard Shapiro and Dr. Rich Wilson, and tentatively decided to sequence two peanut cultivars and their 100 recombinant inbred line (RIL) progenies in collaboration with Chinese peanut researchers. Nwosu and Guo made another trip to China from March 18 to 31, 2011 to discuss the technical strategies and cost-share with Chinese collaborators, Zhang Xinyou, Han Suoyi, Ma Wenyue, Wang Xingjun, Huang Jiaquan, Tang Ronghua, and Liang Xuanqiang and Chen Xiaoping (on conference call). The group unanimously agreed to join the peanut genome sequencing projects, and signed a Memorandum of Understanding.

A third trip was made by US delegation to China (September 19 to 30, 2011) to discuss the time for launching the peanut genome sequencing project with Chinese collaborators and China Ministry of Science and Technology. The members are Victor Nwosu, Kim Moore, Howard Valentine, and Baozhu Guo. The sequencing and assembly strategies as proposed by BGI (Beijing Genome Institute), are adopting an integrated strategy WGS plus BAC by BAC sequencing with Hiseq2000 technology, and then resequencing and calling SNPs according to alignment to the reference genome with 200 RILs. The SNPs will be used as markers to construct a genetic map for chromosome-level assembly. Detailed information will be posted onhttp://www.peanutbioscience.com/ and forthcoming project website and other means.

NEXT-GENERATION SEQUENCING TECHNOLOGIES AND EST PROJECT

In recent years, we are experiencing next-generation sequencing (NGS) technologies which are bringing a revolution in both plant and animal genomics research because of continuous developing in technology resulting in continuous decline in sequencing cost and continuous increase in length of sequence read. A number of NGS technologies such as 454, Illumina, and SOLiD have recently become available. These technologies are capable of generating hundreds of thousands or tens of millions of short DNA sequence reads at a relatively low cost. These NGS technologies are being utilized for de novo sequencing, genome re-sequencing, and whole genome and transcriptome analysis. New generation of sequencers, like the Single-Molecule Real-Time (SMRT) Sequencer, HeliScope Single Molecule Sequencer, and the Personal Genome Machine (PGM) are becoming available that are capable of generating longer sequence reads in a shorter time and at even lower cost for sequencing [75].

Despite the completion of several plant whole genome sequences, much of these genomic data are not fully or well understood. A complete genome annotation requires information and knowledge of the whole genome structure, such as transcription start and stop sites, exon-intron structures, splice variants, and regulatory region sequences. Therefore, EST project will continue to play an important role in postgenome sequencing and will apply NGS technologies in transcriptome sequencing. Sanger-based transcriptome sequencing in the form of ESTs has provided an accurate and effective means for annotating many of the more abundant protein-coding genes [76, 77]. However, the limitations of the Sanger sequencing method restrict the utility of these approaches to the annotation of most abundantly expressed genes. For instance, it has been estimated that most EST studies using Sanger sequencing detect only about 60% of transcripts in the cell and thus do not provide a complete representation of the transcriptome [78]. This information gap can be addressed using the next-generation sequencing technologies. For instance, in peanut, Zhuang et al. [39] generated more than 2 million short sequences using 454 sequencing technology, and Zhao et al. [79] used Higseq2000 producing about 14 million short sequences which could form more than 70 thousand unigenes compared to 1,345 ESTs generated by Sanger sequencing in an earlier study [5, 35, 36].

In genome annotation studies, ESTs are aligned to reference genome sequences, thus revealing the presence of exons, introns, exon junctions, and transcription boundaries for the captured genes. The 454 technology has been used to provide annotation information for the genome of maize [80]. NGS applied in transcriptome sequences has achieved sufficient sequencing depth and provides an adequate representation of the cellular transcriptome. NGS technology for peanut transcriptome analysis has been employed to generate a large number of ESTs and unigenes from different tissue and developmental stages of peanut seeds [39, 79]. The huge numbers of ESTs generated by high throughput sequencing technologies will greatly facilitate peanut functional genomic studies.

CONCLUSION

Millions of EST sequences from several hundred plant species have been deposited in publicly available expressed sequence tag (EST) databases. Many of these ESTs have been sequenced as an alternative to complete genome sequencing or as a substrate for cDNA array-based expression analyses. Although EST collections are certainly no substitute for a whole genome scaffold, this resource forms the core foundations for various genome-scale experiments before and even after genome sequencing.

The peanut research community has reached the level of 252,832 ESTs in the public NCBI EST database (March 2012). The EST sequence resources have been shown to have a wide range of applications and novel uses. There are two historic events, March 2004 Atlanta Workshop deciding that EST was the priority of peanut genome research and December 2010 the Peanut Genome Project Inaugural Meeting deliberating to sequence peanut whole genomes. The ESTs will continue to be actively sequenced to fill in knowledge gaps and complement to whole genome sequences and annotation. EST sequencing certainly played a key role during the pre-era of whole genome sequence and will continue providing an excellent resource for novel gene discovery and for confirmation of in silico gene predictions from genomic DNA sequences. One of the main challenges of low level of genetic diversity has been partially tackled by developing several thousand EST-SSR markers for peanut. Peanut genomic resources continue to build large EST data sets through NGS technologies and thousands of SNP and SSR markers will be developed. Further, it would be possible to integrate the genetic and physical maps in order to facilitate gene cloning and molecular breeding effectively.

ACKNOWLEDGMENTS

The authors thank Billy Wilson for technical assistance in the field and the laboratory. This research was partially supported by funds provided by the USDA Agricultural Research Service and the Georgia Agricultural Commodity Commission for Peanut and Peanut Foundation. Mention of trade names or commercial products in this publication is solely for the purpose of providing specific information and does not imply recommendation or endorsement by the US Department of Agriculture. USDA is an equal opportunity provider and employer.

REFERENCES

1. J. J. Doyle, "Leguminosae," in Encyclopedia of Genetics, S. Brenner and J. H. Miller, Eds., pp. 1081–1085, Academic Press, San Diego, Calif, USA, 2001.
2. B. Z. Guo, C. Y. Chen, Y. Chu, C. C. Holbrook, P. Ozias-Akins, and H. T. Stalker, "Advances in genetics and genomics for sustainable peanut production," in Sustainable Agriculture and New Biotechnologies, N. Benkeblia, Ed., pp. 341–367, CRC Press, Boca Raton, Fla, USA, 2012.
3. R. F. Wilson, H. T. Stalker, and E. C. Brummer, Legume Crop Genomics, AOCS Press, Champaign, Ill, USA, 2004.
4. M. D. Adams, J. M. Kelley, J. D. Gocayne et al., "Complementary DNA sequencing: expressed sequence tags and human genome project," Science, vol. 252, no. 5013, pp. 1651–1656, 1991.
5. M. Luo, P. Dang, B. Z. Guo et al., "Generation of expressed sequence tags (ESTS) for gene discovery and marker development in cultivated peanut,"Crop Science, vol. 45, no. 1, pp. 346–353, 2005.
6. K. Proite, S. C. M. Leal-Bertioli, D. J. Bertioli et al., "ESTs from a wild Arachisspecies for gene discovery and marker development," BMC Plant Biology, vol. 7, article 7, 2007.
7. B. Z. Guo, X. P. Chen, P. Dang et al., "Peanut gene expression profiling in developing seeds at different reproduction stages during Aspergillus parasiticus infection," BMC Developmental Biology, vol. 8, article 12, 2008.

8. B. Z. Guo, X. P. Chen, Y. B. Hong et al., "Analysis of gene expression profiles in leaf tissues of cultivated peanuts and development of EST-SSR markers and gene discovery," International Journal of Plant Genomics, vol. 2009, Article ID 715605, 14 pages, 2009.
9. A. Haegeman, J. Jacob, B. Vanholme, T. Kyndt, M. Mitreva, and G. Gheysen, "Expressed sequence tags of the peanut pod nematode Ditylenchus africanus: the first transcriptome analysis of an Anguinid nematode," Molecular and Biochemical Parasitology, vol. 167, no. 1, pp. 32–40, 2009.
10. Y. P. Bi, W. Liu, H. Xia et al., "EST sequencing and gene expression profiling of cultivated peanut (Arachis hypogaea L.)," Genome, vol. 53, no. 10, pp. 832–839, 2010.
11. G. Q. Song, M. J. Li, H. Xiao et al., "EST sequencing and SSR marker development from cultivated peanut (Arachis hypogaea L.)," Electronic Journal of Biotechnology, vol. 13, no. 3, pp. 1–9, 2010.
12. S. V. Tirumalaraju, M. Jain, and M. Gallo, "Differential gene expression in roots of nematode-resistant and -susceptible peanut (Arachis hypogaea) cultivars in response to early stages of peanut root-knot nematode (Meloidogyne arenaria) parasitization," Journal of Plant Physiology, vol. 168, no. 5, pp. 481–492, 2011.
13. C. C. Holbrook and A. K. Culbreath, "Registration of "Tifrunner" peanut,"Journal of Plant Registrations, vol. 1, article 124, 2007.
14. D. M. T. Alves, R. W. Pereira, S. C. M. Leal-Bertioli, M. C. Moretzsohn, P. M. Guimarães, and D. J. Bertioli, "Development and use of single nucleotide polymorphism markers for candidate resistance genes in wild peanuts (Arachis spp)," Genetics and Molecular Research, vol. 7, no. 3, pp. 631–642, 2008.
15. P. Koilkonda, S. Sato, S. Tabata et al., "Large-scale development of expressed sequence tag-derived simple sequence repeat markers and diversity analysis in Arachis spp.," Molecular Breeding, vol. 30, no. 1, pp. 125–138, 2012.
16. Y. S. Yan, X. D. Lin, Y. S. Zhang, L. Wang, K. Wu, and S. Z. Huang, "Isolation of peanut genes encoding arachins and conglutins by expressed sequence tags," Plant Science, vol. 169, no. 2, pp. 439–445, 2005.

17. L. Wang, Y. S. Yan, B. Liao, X. D. Lin, and S. Z. Huang, "The cDNA cloning of conarachin gene and its expression in developing peanut seeds," Journal of Plant Physiology and Molecular Biology, vol. 31, no. 1, pp. 107–110, 2005.
18. X. J. Wang, L. Su, X. Q. Quan, L. Shan, H. T. Zhang, and Y. P. Bi, "Peanut (Arachis hypogaea L.) EST sequencing, gene cloning and Agrobacteria-mediated transformation," in Proceedings of the International Groundnut Conference on Groundnut Aflatoxin and Genomics, pp. 59–60, Guangzhou, China, November, 2006.
19. J. Y. Wang, L. J. Pan, Q. L. Yang, and S. L. Yu, "Development and characterization of EST-SSR makers from NCBI and cDNA library in cultivated peanut (Arachis hypogaea L.)," Molecular Plant Breeding, vol. 7, no. 4, pp. 806–810, 2009.
20. J. Y. Wang, L. J. Pan, Q. L. Yang, and S. L. Yu, "Development and characterization of EST-SSR markers from NCBI and cDNA library in cultivated peanut (Arachis hypogaea L.)," Legume Genomics and Genetics, vol. 1, no. 6, pp. 30–33, 2010.
21. J. Q. Huang, L. Y. Yan, Y. Lei, H. F. Jiang, and B. S. Liao, "Peanut cDNA library construction and EST sequence analysis," Chinese Journal of Oil Crop Sciences, vol. 10, pp. 121–125, 2008.
22. Y. Xiao, B. S. Liao, L. Y. Yan et al., "Development and utilization of EST-SSR primers in peanut(Arachis hypogaea L.)," Hubei Agricultural Sciences, vol. 49, no. 11, pp. 2625–2628, 2010.
23. M. J. Li, H. Xia, C. Z. Zhao et al., "Isolation and characterization of putative Acetyl-CoA carboxylases in Arachis hypogaea L.," Plant Molecular Biology Reporter, vol. 28, no. 1, pp. 58–68, 2009.
24. M. J. Li, A. Q. Li, H. Xia et al., "Cloning and sequence analysis of putative type II fatty acid synthase genes from Arachis hypogaea L.," Journal of Biosciences, vol. 34, no. 2, pp. 227–238, 2009.
25. M. J. Li, X. J. Wang, L. Su, Y. P. Bi, and S. B. Wan, "Characterization of five putative acyl carrier protein (ACP) isoforms from developing seeds of Arachis hypogaea L.," Plant Molecular Biology Reporter, vol. 28, no. 3, pp. 365–372, 2010.
26. R. M. P. Siloto, K. Findlay, A. Lopez-Villalobos, E. C. Yeung, C. L. Nykiforuk, and M. M. Moloney, "The accumulation of oleosins determines the size of seed oilbodies in Arabidopsis," Plant Cell, vol. 18, no. 8, pp. 1961–1974, 2006.

27. X. J. Wang, H. Xia, C. S. Li, C. Z. Zhao, and A. Q. Li, Peanut Allergens: Gene Cloning and RNAi Interference. Sino-Dutch Symposium on Multedisciplinary Allergy Research, Hangzhou, China.
28. L. Su, C. Z. Zhao, Y. P. Bi, S. B. Wan, H. Xia, and X. J. Wang, "Cloning and expression analysis of peanut LEA protein genes," Journal of Biosciences, vol. 36, no. 2, pp. 223–228, 2011.
29. F. X. Shao, Z. J. Liu, L. Q. Wei, M. Cao, and Y. P. Bi, "Cloning and sequence analysis of a novel NAC-like gene AhNAC1 in Peanut (Arachis hypogaea),"Acta Botanica Boreali-Occidentalia Sinica, vol. 10, pp. 1929–1934, 2008.
30. M. Zhang, W. Liu, Y. P. Bi, and Z. Z. Wang, "Isolation and identification of PNDREB1: a new DREB transcription factor from peanut (Arachis hypogaeaL.)," Acta Agronomica Sinica, vol. 35, no. 11, pp. 1973–1980, 2009.
31. G. L. Zhang, X. G. Shi, N. B. Cai, Y. L. Zhao, and W. J. Zhuang, "Molecular cloning and expression of pericarp and testaspecific expression gene AhPSG13," Chinese Journal of Oil Crop Sciences, vol. 32, no. 1, pp. 035–040, 2010.
32. W. F. Li, H. Y. Liu, N. Zhong, and X. Q. Liang, "Cloning and prokaryotic expression of defensin gene from peanut (Arachis hypogaea L.)," Genomics and Applied Biology, vol. 28, no. 4, pp. 645–650, 2009.
33. C. Z. Xie, X. Q. Liang, L. Li, and H. Y. Liu, "Cloning and prokaryotic expression of AhPR10 gene with resistance to aspergillus flavus in peanut,"Genomics and Applied Biology, vol. 28, no. 2, pp. 237–244, 2009.
34. C. Z. Zhao, A. Q. Li, X. J. Wang, H. Xia, L. Su, and C. S. Li, "Cloning and expression analysis of lipid transfer protein family genes in Arachis hypogaeaL.," Journal of Peanut Science, vol. 38, no. 4, pp. 15–20, 2009.
35. M. Luo, P. Dang, M. G. Bausher et al., "Identification of transcripts involved in resistance responses to leaf spot disease caused by Cercosporidium personatum in peanut (Arachis hypogaea)," Phytopathology, vol. 95, no. 4, pp. 381–387, 2005.
36. M. Luo, X. Liang, P. Dang et al., "Microarray-based screening of differentially expressed genes in peanut in response to Aspergillus

parasiticus infection and drought stress," Plant Science, vol. 169, no. 4, pp. 695–703, 2005.

37. P. Payton, K. R. Kottapalli, D. Rowland et al., "Gene expression profiling in peanut using high density oligonucleotide microarrays," BMC Genomics, vol. 10, article 265, 2009.
38. B. Z. Guo, N. D. Fedorova, X. P. Chen et al., "Gene expression profiling and identification of resistance genes to Aspergillus flavus infection in peanut through EST and microarray strategies," Toxins, vol. 3, no. 7, pp. 737–753, 2011.
39. W. J. Zhuang, H. Chen, P. K. Nancy et al., "Isolation and characterization of important genes toward improvement peanut resistance to Aspergillus flavus," in Proceedings of the 5th International Conference of the Peanut Research Community on Advances in Arachis Through Genomics and Biotechnology, p. 28, Brasillia, Brazil, June, 2011.
40. X. Q. Liang, X. P. Chen, Y. B. Hong et al., "Utility of EST-derived SSR in cultivated peanut (Arachis hypogaea L.) and Arachis wild species," BMC Plant Biology, vol. 9, article 35, 2009.
41. H. Qin, S. Feng, C. Chen et al., "An Integrated genetic linkage map of cultivated peanut (Arachis hypogaea L.) constructed from two RIL populations," Theoretical and Applied Genetics, vol. 124, pp. 653–664, 2012.
42. X. Q. Liang, Y. B. Hong, X. P. Chen et al., "Characterization and application of EST-SSRs in peanut (Arachis hypogaea L.)," Acta Agronomica Sinica, vol. 35, no. 2, pp. 246–254, 2009.
43. T. Thiel, W. Michalek, R. K. Varshney, and A. Graner, "Exploiting EST databases for the development and characterization of gene-derived SSR-markers in barley (Hordeum vulgare L.)," Theoretical and Applied Genetics, vol. 106, no. 3, pp. 411–422, 2003.
44. C. Liewlaksaneeyanawin, C. E. Ritland, Y. A. El-Kassaby, and K. Ritland, "Single-copy, species-transferable microsatellite markers developed from loblolly pine ESTs," Theoretical and Applied Genetics, vol. 109, no. 2, pp. 361–369, 2004.
45. R. K. Varshney, R. Sigmund, A. Börner et al., "Interspecific transferability and comparative mapping of barley EST-SSR markers in wheat, rye and rice,"Plant Science, vol. 168, no. 1, pp. 195–202, 2005.

46. S. P. Feng, W. G. Li, H. S. Huang, J. Y. Wang, and Y. T. Wu, "Development, characterization and cross-species/genera transferability of EST-SSR markers for rubber tree (Hevea brasiliensis)," Molecular Breeding, vol. 23, no. 1, pp. 85–97, 2009.
47. B. S. Bushman, S. R. Larson, M. Tuna et al., "Orchardgrass (Dactylis glomerata L.) EST and SSR marker development, annotation, and transferability," Theoretical and Applied Genetics, vol. 123, no. 1, pp. 119–129, 2011.
48. S. Choudhary, N. K. Sethy, B. Shokeen, and S. Bhatia, "Development of chickpea EST-SSR markers and analysis of allelic variation across related species," Theoretical and Applied Genetics, vol. 118, no. 3, pp. 591–608, 2009.
49. R. Koppolu, H. D. Upadhyaya, S. L. Dwivedi, D. A. Hoisington, and R. K. Varshney, "Genetic relationships among seven sections of genus Arachisstudied by using SSR markers," BMC Plant Biology, vol. 10, article 15, 2010.
50. E. S. Mace, R. K. Varshney, V. Mahalakshmi et al., "In silico development of simple sequence repeat markers within the aeschynomenoid/dalbergoid and genistoid clades of the Leguminosae family and their transferability to Arachis hypogaea, groundnut," Plant Science, vol. 174, no. 1, pp. 51–60, 2008.
51. R. K. Varshney, D. J. Bertioli, M. C. Moretzsohn et al., "The first SSR-based genetic linkage map for cultivated groundnut (Arachis hypogaea L.),"Theoretical and Applied Genetics, vol. 118, no. 4, pp. 729–739, 2009.
52. M. C. Moretzsohn, L. Leoi, K. Proite et al., "A microsatellite-based, gene-rich linkage map for the AA genome of Arachis (Fabaceae)," Theoretical and Applied Genetics, vol. 111, no. 6, pp. 1060–1071, 2005.
53. M. C. Moretzsohn, A. V. G. Barbosa, D. M. T. Alves-Freitas et al., "A linkage map for the B-genome of Arachis (Fabaceae) and its synteny to the A-genome," BMC Plant Biology, vol. 9, article 40, 2009.
54. D. Foncéka, T. Hodo-Abalo, R. Rivallan et al., "Genetic mapping of wild introgressions into cultivated peanut: a way toward enlarging the genetic basis of a recent allotetraploid," BMC Plant Biology, vol. 9, article 103, 2009.

55. S. C. M. Leal-Bertioli, A. C. V. F. José, D. M. T. Alves-Freitas et al., "Identification of candidate genome regions controlling disease resistance inArachis," BMC Plant Biology, vol. 9, article 112, 2009.
56. K. Ravi, V. Vadez, S. Isobe et al., "Identification of several small main-effect QTLs and a large number of epistatic QTLs for drought tolerance related traits in groundnut (Arachis hypogaea L.)," Theoretical and Applied Genetics, vol. 122, no. 6, pp. 1119–1132, 2011.
57. Y. B. Hong, X. P. Chen, X. Q. Liang et al., "A SSR-based composite genetic linkage map for the cultivated peanut (Arachis hypogaea L.) genome," BMC Plant Biology, vol. 10, article 17, 2010.
58. S. Mondal, A. M. Badigannavar, and S. F. D'Souza, "Molecular tagging of a rust resistance gene in cultivated groundnut (Arachis hypogaea L.) introgressed from Arachis cardenasii," Molecular Breeding, vol. 29, no. 2, pp. 467–476, 2012.
59. D. A. Knauft and D. W. Gorbet, "SunOleic 97R," Florida Agricultural Experiment Stations Circular, vol. 400, pp. 1–7, 1997.
60. D. W. Gorbet and D. A. Knauft, "Registration of "SunOleic 95R" peanut,"Crop Science, vol. 4, article 1932, 1997.
61. A. J. Norden, D. W. Gorbet, and D. A. Knauft, "Registration of "Sunrunner" peanut," Crop Science, vol. 25, article 1126, 1985.
62. A. K. Culbreath, D. W. Gorbet, N. Martinez-Ochoa et al., "High levels of field resistance to tomato spotted wilt virus in peanut breeding lines derived from hypogaea and hirsuta botanical varieties," Peanut Science, vol. 32, pp. 20–24, 2005.
63. Y. Chu, L. Ramos, C. C. Holbrook, and P. Ozias-Akins, "Frequency of a loss-of-function mutation in oleoyl-PC desaturase (ahFAD2A) in the mini-core of the U.S. peanut germplasm collection," Crop Science, vol. 47, no. 6, pp. 2372–2378, 2007.
64. Y. Chu, C. C. Holbrook, and P. Ozias-Akins, "Two alleles of ahFAD2B control the high oleic acid trait in cultivated peanut," Crop Science, vol. 49, no. 6, pp. 2029–2036, 2009.
65. G. Kochert, T. Halward, W. D. Branch, and C. E. Simpson, "RFLP variability in peanut (Arachis hypogaea L.) cultivars and wild

species," Theoretical and Applied Genetics, vol. 81, no. 5, pp. 565–570, 1991.

66. G. He and C. Prakash, "Evaluation of genetic relationships among botanical varieties of cultivated peanut (Arachis hypogaea L.) using AFLP markers,"Genetic Resources and Crop Evolution, vol. 48, no. 4, pp. 347–352, 2001.

67. T. Halward, H. T. Stalker, and G. Kochert, "Development of an RFLP linkage map in diploid peanut species," Theoretical and Applied Genetics, vol. 87, no. 3, pp. 379–384, 1993.

68. M. D. Burow, C. E. Simpson, J. L. Starr, and A. H. Paterson, "Transmission genetics of chromatin from a synthetic amphidiploid to cultivated peanut (Arachis hypogaea L.): broadening the gene pool of a monophyletic polyploid species," Genetics, vol. 159, no. 2, pp. 823–837, 2001.

69. H. T. Stalker and J. P. Moss, "Speciation, Cytogenetics, and Utilization ofArachis Species," Advances in Agronomy, vol. 41, no. C, pp. 1–40, 1987.

70. G. Kochert, H. T. Stalker, M. Gimenes, L. Galgaro, C. R. Lopes, and K. Moore, "RFLP and cytogenetic evidence on the origin and evolution of allotetraploid domesticated peanut, Arachis hypogaea (Leguminosae)," American Journal of Botany, vol. 83, no. 10, pp. 1282–1291, 1996.

71. N. D. Young, N. F. Weeden, and G. Kochert, "Genome mapping in legumes (Fam. Fabaceae)," in Genome Mapping in Plants, A. H. Paterson, Ed., pp. 211–227, Landes Bioscience Press, Austin, Tex, USA, 1996.

72. S. A. Goff, D. Ricke, T. H. Lan et al., "A draft sequence of the rice genome (Oryza sativa L. ssp. japonica)," Science, vol. 296, no. 5565, pp. 92–100, 2002.

73. J. Yu, S. Hu, J. Wang, G. K. Wong, S. Li, B. Liu, et al., "A draft sequence of the rice genome (Oryza sativa L. ssp. indica)," Science, vol. 296, no. 5565, pp. 79–92, 2002.

74. A. G. Tian, J. Wang, P. Cui et al., "Characterization of soybean genomic features by analysis of its expressed sequence tags," Theoretical and Applied Genetics, vol. 108, no. 5, pp. 903–913, 2004.

75. M. Thudi, Y. Li, S. A. Jackson, G. D. May, and R. K. Varshney, "Current state-of-art of sequencing technologies for plant genomics research," Briefings in Functional Genomics, vol. 11, pp. 3–11, 2012.
76. L. D. Hillier, G. Lennon, M. Becker et al., "Generation and analysis of 280,000 human expressed sequence tags," Genome Research, vol. 6, no. 9, pp. 807–828, 1996.
77. M. Seki, M. Narusaka, A. Kamiya et al., "Functional annotation of a full-length Arabidopsis cDNA collection," Science, vol. 296, no. 5565, pp. 141–145, 2002.
78. M. R. Brent, "Steady progress and recent breakthroughs in the accuracy of automated genome annotation," Nature Reviews Genetics, vol. 9, no. 1, pp. 62–73, 2008.
79. C. Z. Zhao, X. J. Wang, A. Q. Li, and C. S. Li, "De novo characterization of peanut transcriptome during gynophore development," in Proceedings of the 5th International Conference of the Peanut Research Community on Advances in Arachis Through Genomics and Biotechnology, p. 48, Brasillia, Brazil, June, 2011.
80. S. J. Emrich, W. B. Barbazuk, L. Li, and P. S. Schnable, "Gene discovery and annotation using LCM-454 transcriptome sequencing," Genome Research, vol. 17, no. 1, pp. 69–73, 2007.

Chapter 7

Groundnut Improvement: Use of Genetic and Genomic Tools

Pasupuleti Janila[1], S. N. Nigam[1] Manish K. Pandey[1] P. Nagesh[1] and Rajeev K. Varshney[1,2]

[1]International Crops Research Institute for the Semi-Arid Tropics, Patancheru, Andhra Pradesh, India

[2]Generation Challenge Programme, c/o Centro Internacional de Mejoramiento de Maíz y Trigo, Mexico DF, Mexico

ABSTRACT

Groundnut (*Arachis hypogaea* L.), a self-pollinated legume is an important crop cultivated in 24 million ha world over for extraction of edible oil and food uses. The kernels are rich in oil (48–50%) and protein (25–28%), and are source of several vitamins, minerals, antioxidants, biologically active polyphenols, flavonoids, and isoflavones. Improved varieties of groundnut with high yield potential were developed and released for cultivation world over. The improved varieties belong to different maturity durations and possess resistance to diseases, tolerance to drought, enhanced oil content, and improved quality traits for food uses. Conventional breeding procedures along with the tools for phenotyping were largely used in groundnut improvement programs. Mutations were used to induce variability and wide hybridization was attempted to tap variability from wild species. Low genetic variability has been a bottleneck for groundnut improvement. The vast potential of wild species, reservoir of new alleles remains under-utilized. Development of linkage maps of groundnut during the last decade was followed by identification of markers and quantitative trait loci for the target traits. Consequently, the last decade has witnessed the deployment of molecular breeding approaches to complement the ongoing groundnut improvement programs in USA, China, India, and Japan. The other potential advantages of molecular breeding are the feasibility to target multiple traits for improvement and provide tools to tap new alleles from wild species. The first groundnut variety developed through marker-assisted back-crossing is a root-knot nematode-resistant variety, NemaTAM in USA. The uptake of molecular breeding approaches in groundnut improvement programs by NARS partners in India and many African countries is slow or needs to be initiated in part due to inadequate infrastructure, high genotyping costs, and human capacities. Availability of draft genome sequence for diploid (AA and BB) and tetraploid, AABB genome species of *Arachis* in coming years is expected to bring low-cost genotyping to the groundnut community that will facilitate use of modern genetics and breeding approaches such as genome-wide association studies for trait mapping and genomic selection for crop improvement.

INTRODUCTION

Economic Importance and Uses

Groundnut, also known as peanut, is an important oil, food, and feed legume crop grown in over 100 countries. It covered 24 million ha area worldwide with a total production of 38 million tons in 2010 (FAOSTAT, 2010). In the last decade (2000–2010), the global groundnut production increased marginally. The global annual increase in production was 0.4% which was due to both, an annual increase in yield by 0.1% and in area by 0.3% (Figure1). The projected demand of groundnut in Asia alone by 2020 is expected to be 1.6 times more than the level of production in 2000 (Birthal et al., 2010). If the projected demands have to be met, the productivity and production of the crop has to increase at a much higher growth rate than the present one. Asia and Africa account for 95% of global groundnut area where it is cultivated under rainfed conditions with low inputs by resource poor farmers. Groundnut is a cash crop providing income and livelihoods to the farmer. It also contributes to nutrition of farm families through consumption of energy- and protein-rich groundnut kernels and provides nutritious fodder (haulms) to livestock. Thus groundnut cultivation contributes to the sustainability to mixed crop-livestock production systems, the most predominant system of the semi-arid areas.

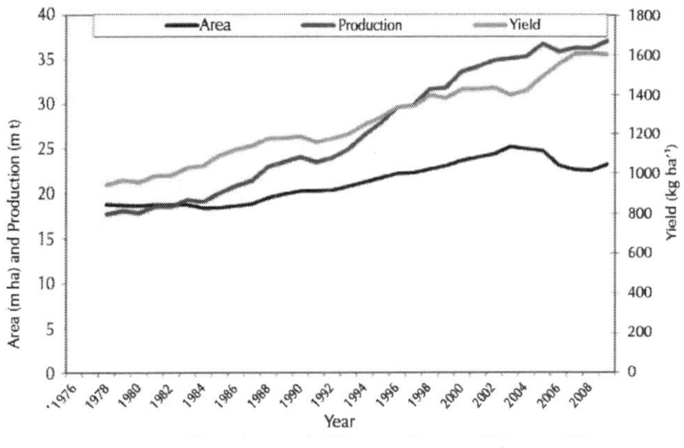

Figure 1. Three-year moving center average for groundnut pod yield, production, and area harvested in world.

Groundnut is valued as a rich source of energy contributed by oil (48–50%) and protein (25–28%) in the kernels. They provide 564 kcal of energy from 100 g of kernels (Jambunathan, 1991). In addition, the groundnut kernels contain many health enhancing nutrients such as minerals, antioxidants, and vitamins and are rich in monounsaturated fatty acids. They contain antioxidants like *p*-coumaric acid and resveratrol, Vitamin E, and many important B-complex groups of thiamin, pantothenic acid, vitamin B-6, folates, and niacin. Groundnut is a dietary source of biologically active polyphenols, flavonoids, and isoflavones. As they are highly nutritious, groundnut and products based on groundnut can be promoted as nutritional foods to fight energy, protein, and micronutrient malnutrition among the poor. Groundnut-based Plumpy'nut[1], a ready to use therapeutic food, has helped save the lives of thousands of malnourished children in Niger (UNICEF, 2007).

Over 60% of global groundnut production is crushed for extraction of oil for edible and industrial uses, while 40% is consumed in food uses and others (such as seed for sowing the next season crop; Birthal et al., 2010). Groundnut oil is an excellent cooking medium because of its high smoking point (Singh and Diwakar, 1993). India, China, Myanmar, and Vietnam use groundnut oil for cooking purpose extensively. The cake obtained after extraction of oil is used in animal feed industry, in preparing enriched easily digestible food for children and aged persons, and as soil amendment. In Europe and North and South America about 75% of the production is used as food, while only 35% is used for the same purpose in Asia (Birthal et al., 2010). Peanut butter is the most popular groundnut product in the USA, Canada, and Australia. Groundnut seed can be consumed raw (non-heated), boiled, and roasted and also used to make confections and its flour to make baked products. The groundnut shells are used for making particle boards or used as fuel or filler in fertilizer and feed industry. Groundnut haulms constitute nutritious fodder for livestock. They contain protein (8–15%), lipids (1–3%), minerals (9–17%), and carbohydrate (38–45%) at levels higher than cereal fodder. The digestibility of nutrients in groundnut haulm is around 53% and that of crude protein 88% when fed to cattle. Haulms release energy up to 2337 cal kg^{-1} of dry matter. Being a legume crop, groundnut helps in improving soil health and fertility by leaving behind N_2 and organic matter in the soil.

Taxonomy and Biology

The cultivated groundnut (*Arachis hypogaea* L.), an annual herb belonging to the family Fabaceae (Leguminosae), is classified into two subspecies, subsp. *fastigiata* Waldron and subsp. *hypogaea* Krap.

et. Rig. The subsp. *fastigiata* contains four botanical varieties, var. *vulgaris*, var. *fastigiata*, var. *peruviana*, and var. *aequatoriana*. The subsp. *hypogaea* contains two varieties, var. *hypogaea* and var. *hirsuta*. Each of these botanical types has different plant, pod, and seed characteristics (Krapovickas and Gregory, 1994). Groundnut is an allotetraploid (2n = 2x = 40) with "AA" and "BB" genomes. All species, except the cultivated species (*A. hypogaea* and *A. monticola*) in Section *Arachis*, and certain species in Section*Rhizomatosae*, are diploid (2n = 2x = 20). The diploid progenitors, *A. duranensis* and *A. ipaensis*, contributed "AA" and "BB" genomes, respectively, to the cultivated groundnut (Kochert et al., 1996). The phylogenetic analyses based on intron sequences and microsatellite markers also provide evidence for this hypothesis (Moretzsohn et al., 2012). A single hybridization event between the diploid progenitors followed by chromosome doubling (Kochert et al., 1996) about 3500 years ago lead to origin of cultivated groundnut. Southern Bolivia and Northern Argentina are thought to be center of origin of this crop (Gregory et al., 1980;Kochert et al., 1996). The center of diversity of the genus includes Western Brazil, Bolivia, Paraguay, and Northern Argentina (Gregory et al., 1980). *A. duranensis* occurs throughout the region, while *A. ipaensis* has only been found in Southern Bolivia. The genetic diversity of the genus is classified into four gene pools (Singh and Simpson, 1994): primary gene pool consisting of *A. hypogaea* and *A. monticola*, secondary consisting of diploid species from Section *Arachis* that are cross-compatible with *A. hypogaea*, tertiary consisting of species of the Section Procumbentes that are weakly cross-compatible with *A. hypogaea*, and the fourth gene pool consisting of the remaining wild *Arachis* species classified into seven other sections.

Groundnut is a self-pollinated crop with cleistogamous flowers, but natural hybridization can occur to small extent where bee activity is high (Nigam et al., 1983). Flowering begins 17–35 days after seedling emergence depending on the cultivar and environmental conditions. Flowers, simple or compound, are born in the axils of leaves and never at the same node as vegetative branch. One or more flowers may be present at a node. The stigma becomes receptive to pollen about 24-h before anthesis and remains so for about 12 h after anthesis, and the dehiscence of anthers takes place 2–3 h prior to opening of the flower in the morning. Fertilization occurs about 6 h after pollination. Depending upon the prevailing temperatures, the peg or gynophore carrying the ovary and fertilized ovule on its tip appears in 6–10 days and grows to enter the soil (positively geotropic) where it develops into pods. The tip orients itself horizontally away from tap root. Groundnut grows well in well-distributed rainfall of at least 500 mm. The growth and development is largely influenced by temperature in groundnut and the optimum air temperature is between 25 and 30°C. The nutritional requirement of groundnut is different as the pods develop in the soil. Calcium is an important nutrient required for pod and kernel development. It is unique to groundnuts that the pods directly absorb most of the calcium, and therefore calcium fertilizers are applied in the pod zone at the peak flowering stage to ensure its availability to the pods.

TARGET TRAITS FOR GROUNDNUT IMPROVEMENT

The aim of groundnut breeding programs across the world is to develop new varieties that meet the requirements of grower, processor, and consumer. Thus the targeted traits for improvement in groundnut depend on the level of productivity achieved and consumers' and industry requirements in a country. In the USA, where average productivity is high, the focus is more on improving food quality and flavor traits and freedom from mycotoxins in the produce. In the developing countries, where yields are generally low, the focus is more on removing yield barriers besides improving yield *per se*.

Yield Parameters and Adaptation

Yield and yield contributing parameters are the most widely targeted traits of groundnut improvement programs worldwide. Selection for yield *per se* has been the major basis for improving groundnut productivity in the world (Nigam et al., 1991), but gains from such selection have been low and slow due to large g × e interactions observed for these traits. The pod yield is a function of crop growth rate, duration of reproductive growth, and the fraction of crop growth rate partitioned toward pod yield. Therefore understanding physiology of yield is also essential to better target yield increase. The important yield contributing parameters are: pod yield per plant, number of pods per plant, shelling outturn, and 100-seed weight. Recently, physiological traits associated with yield (harvest index, transpiration-use-efficiency, etc.) are also receiving attention in breeding programs which have necessary infrastructure and resources. Other traits not having direct bearing on yield such as ease in shelling, ease in harvesting (peg strength), number of seeds per pod (for specific uses), reticulation, beak and constriction of pod, kernel shape and color, and blanching ability are also important considerations to satisfy farmers', processors', and market demands. For the development of dual purpose varieties, haulm yield becomes an important consideration in addition to pod yield. In addition to quantity of haulm produced, its quality determined by nitrogen content and digestibility are also important to breed dual purpose varieties. Crop duration and fresh seed dormancy in Spanish varieties are important adaptation traits. The maturity duration should match with the length of growing period (LGP; 90 to over 150 days) available at a given location and conditioned by the soil moisture availability and climatic conditions (mainly temperatures and sunshine hours). Early maturity is an important trait in groundnut as it enables escape from stress conditions such as drought and frost and to fit in multiple cropping systems. *In situ* germination, a consequence of lack of fresh seed dormancy leads to pod yield and quality loss in rainfed environments, particularly when rains coincide with the crop maturity stage.

Stress Tolerance/Resistance

There is large gap between potential pod yield and the realized pod yield in most of the situations (Johansen and Nageswara Rao, 1996). Potential yield is defined as the maximum yield obtainable by the best genotypes in a specified agro-climatic environment when the known biotic and abiotic constraints are overcome. The yield gap in the groundnut grown under water limiting conditions in rainfed areas is further aggravated by incidence of a host of diseases and insect pests. Therefore, tolerance/resistance traits that offer protection against losses caused by biotic and abiotic stresses are important target traits. In addition to protection to yield, resistance/tolerance to stress factors enhances the quality (nutritional, visual appearance, sensory attributes, free from toxins, and post-harvest keeping quality) of both, pods and haulms that fetches better price in the market. However, studies have shown that high yield potential and high degree of resistance do not generally go together (Nigam et al., 1991), while the breeding programs target them together. Therefore, in most of the breeding programs a balance is struck between the yield potential and level of resistance to avoid any possible yield penalty. As a consequence, several varieties with high yield potential and moderate levels of resistance were bred and released for cultivation world over (see Some Accomplishments of Conventional Approaches). Drought and high temperature are the most important abiotic stresses that are widespread in groundnut-growing areas. Depending on the time of occurrence, drought can be characterized as early season, mid-season, and end-of-season drought. Mid- and end-of-season droughts are critical as they affect the pod yield and quality. Further, end-of-season drought predisposes pre-harvest *Aspergillus* infection in the field that affects quality of produce. Linked closely with drought is high temperature stress. Two key stages for heat stress in groundnut are: flowering including microsporogenesis (3–6 days before flowering), and fruit-set (Craufurd et al., 2002, 2003). The CGIAR's climate change for agriculture and food security (CCAFS) research has shown that high temperature stress (above 30°C) will be widespread in East and Southern Africa, India, South East Asia, and Northern Latin America, which are important groundnut-growing areas. Thus, effort to breed varieties that can thrive and yield under both, drought and heat stress need to be intensified. With the increase in problem soils (saline and acid) across the cultivated lands of the world, breeding for tolerance to salinity and aluminum toxicity in acid soils are considered important target traits for groundnut improvement in some countries such as China.

Groundnut is attacked by several diseases caused by fungi. Late leaf spot (LLS) caused by *Phaeoisariopsis personata* (Berk. & Curt.) Van Arx, early leaf spot (ELS) caused by *Cercospora arachidicola* Hori and rust caused by *Puccinia arachidis* Spegazzini are among the major foliar fungal diseases worldwide. Aflatoxins are potent carcinogen produced by *Aspergillus* spp. infection in seed forcing several countries to have strict regimes in place on permissible levels of aflatoxins in their imports. *A. flavus* is predominant species in Asia and Africa, while *A. parasiticus* is in the USA. Stem and pod rot, caused by *Sclerotium rolfsii*, is potential threat to groundnut production in many warm, humid areas, especially where irrigated groundnut cultivation is expanding. Groundnut is also a host to several virus diseases, but only a few of them are economically important – groundnut rosette disease (GRD) in Africa, peanut bud necrosis disease (PBND) in India, tomato spotted wilt virus (TSWV) in the USA, peanut stripe potyvirus (PStV) in East and South East Asia, peanut stem necrosis disease (PSND) in pockets in Southern India, and peanut clump virus disease (PCVD) in West Africa (Nigam et al., 2012). Bacterial wilt, caused by *Ralstonia solanacearum,* is predominant among bacterial diseases of groundnut. Globally, nematodes cause 11.8% yield loss in groundnut. The root-knot nematodes, *Meloidogyne* spp. and the lesion nematodes, *Pratylenchus* spp. are important in groundnut (Sharma and McDonald, 1990). Aphids (*Aphis craccivora* Koch), several species of thrips (*Frankliniella schultzei, Thrips palmi,* and *F. fusca*), leaf miner (*Aproaerema modicella*), red hairy caterpillar (*Amsacta albistriga*), jassids (*Empoasca kerri* and *E. fabae*), and *Spodoptera* are the major insect pests in groundnut, among which aphids, thrips, and *Spodoptera* have worldwide distribution and cause serious damage (Wightman and Amin, 1988). In addition to causing yield losses, aphids and thrips are vectors of important virus diseases. Termites, white grubs, and storage pests also cause damage to groundnuts. Groundnut borer or weevil (*Caryedon serratus*) and rust-red flour beetle (*Tribolium castaneum*) are the major storage insect pests in groundnut. In most breeding programs across the world, breeding for resistance to diseases has received more attention than breeding for resistance to insect pest except when they are vector of viral disease. Another important reason for this is the availability of the resistant sources for diseases in cultivated and wild *Arachis* species.

Quality Parameters and Other Traits

Quality includes both, physical and chemical attributes. Nutritional traits include oil, protein, sugar, iron and zinc content, fatty acid profile, and freedom from toxins, while the other quality parameters include visual and sensory attributes (consumer and trader preferred traits) and traits desirable in food/oil processing industries. Breeding for quality parameters in addition to yield enhances the economic returns to the farmers and other stakeholders along the value chain. Studies have shown that high oil content in groundnut is translated into economic benefits to both farmer and millers. Similarly, the produce with desirable traits for confectionary uses fetches higher price in the market because of its export value. Traits impacting on food and oil uses are also important; they include both quality and nutritional parameters. Depending on the nature of use, low oil and high protein contents (for food use), high oil content (for oil use), and high oleic/linoleic fatty acid ratio (for longer shelf-life) are important targeted traits in advanced breeding programs. The traits for confectionary uses in India are: greater proportion of sound mature kernels (SMK), flavor, 100 seed weight exceeding 55 g, >11% of sugar content, >24% of protein content, blanchability (>60%), and low oil content (<45%; Ramanathan, 2004). Seed mass is an important attribute to confectionary quality; however like yield and yield parameters, it is highly influenced by environment.

The taste and sensory attributes of roasted groundnuts are associated with carbohydrate components of the kernel (Pattee et al., 2000). Seed color and shape and flavor are the other important confectionary attributes. Blanchability is removal of testa or seed coat (skin) from raw or roasted groundnuts and this attribute is of economic importance in processed groundnut food products, which include peanut butter, salted groundnuts, candies, and bakery products and groundnut flour. Groundnuts are rich source of several micronutrients, among which iron and zinc contents are important. Enhanced levels of iron and zinc are gaining importance with identification of biofortification (delivery of micronutrients *via* micronutrient-dense crops) as a cost-effective and sustainable approach to fight malnutrition among poor. Development of groundnut cultivars suitable to mechanization is important in developed countries and it is also becoming important in developing countries where human labor for agricultural operation is not available or expensive. Some of the traits for such cultivars include strong pegs, uniform maturity of pods, break resistant pods, non-protruding radicle, and less erect plants with pods near the tap root. Biological nitrogen fixation (BNF) is another important target trait in groundnut that can be improved through both, cultivar selection and *Rhizobium* strain improvement.

GROUNDNUT IMPROVEMENT USING CONVENTIONAL APPROACHES

Genetic Variability and Genetics

The genetic variability in groundnut is low due to origin of the crop through a single hybridization event between two diploid species followed by a chromosome doubling and crossing barriers with wild diploid species (due of ploidy differences). The cultivated groundnut is an allotetraploid, while all wild *Arachis*species are diploid except *A. monticola* and certain species in Section Rhizomatosae. The low genetic variability for the traits of importance and polyploidy nature are a bottleneck to the groundnut improvement. The cultivated accession of *Arachis hypogaea* in the genebanks (repositories of plant genetic resources) and the advanced breeding lines in the breeding programs are the most frequently used sources of variability used as parents in hybridization. Groundnut genetic resources are available in the genebanks of ICRISAT, National Bureau of Plant Genetic Resources (NBPGR), and Directorate of Groundnut Research (DGR) in India; Oil Crops Research Institute of Chinese Academy of Agricultural Sciences and Crops Research Institute of Guangdong Academy of Agricultural Sciences in China; U. S. Department of Agriculture, Texas A&M University, and North Carolina State University in the USA; EMBRAPA - CENARGEN and the Instituto Agronômico de Campinas in Brazil; and Instituto Nacional de Tecnología Agropecuaria (INTA) and Instituto de Botánica del Nordeste (IBONE) in Argentina. Selection of an appropriate source that can be used as a parent in hybridization is challenging for two reasons; first adequate variability should be available to identify a source and second, the amenability of trait for improvement that is determined by the genetic nature of the trait. Much of the variability till remains poorly used in improvement programs. Nigam (2000) reported that only three disease resistant parents (J 11, NC Ac 17090, and PI 259747) appear in the parentage of the cultivars released in India. Ensuring genetic diversity in farmer's fields is another important challenge to the breeder when limited number of sources is known.

Genetic variability for yield and yield attributes, resistance/ tolerance to foliar fungal diseases (Singh et al., 2003), insect pests

(Lynch, 1990), root-knot nematode (Simpson et al., 1993), traits for confectionary uses (Dwivedi and Nigam, 2005), oil content and quality (Upadhyaya et al., 2011), haulm yield and quality traits (Nigam and Blummel, 2010), and several other traits of economic importance were reported in literature (Table1). The germplasm accessions of cultivated groundnut in the gene banks are a vast repository of this variation. Wild *Arachis* species are also reported to possess desirable alleles for several economically important traits such as resistance to fungal and virus diseases, insect pests, and abiotic stress (Dwivedi et al., 2003, 2008; Rao et al., 2003; Kalyani et al., 2007; Nautiyal et al., 2008) and therefore can be potential sources for use in improvement of groundnut (Rao et al., 2003). However, a large proportion of cultivated and wild genetic resources of groundnut are yet to be studied and characterized for targeted and other traits of economic importance (see Pandey et al., 2012b for available collections of genetic resources). The concept of core and min-core collections was suggested to enable handling of the vast collections in gene banks and lay hands on appropriate sources. Core (representing 10% of total collection) and mini-core (representing 10% of core collection) collections were made for the groundnut genetic resources available in USDA/ARS (Holbrook et al., 1993; Holbrook and Dong, 2005), ICRISAT (Upadhyaya et al., 2002, 2003), and China (Jiang et al., 2008, 2010). Holbrook and Isleib (2001) observed that the resistance genes were cluster geographically and accessions with multiple disease resistance were most common in India, Mozambique, and Senegal.

Table 1. List of important traits in groundnut for which variability was reported in literature

Trait	Reference
Agronomic traits	
Yield and yield parameters	Nath and Alam (2002)
Pod characters	Swamy et al. (2006)
Seed dormancy	Wang et al. (2012)
Early maturity	Upadhyaya et al. (2006)
Resistance to diseases and insect pests	
Early leaf spot (*Cercospora arachidicola* Hori)	Singh et al. (1997), Holbrook and Isleib (2001)

Late leaf spot [*Phaeoisariopsis personata* (Berk. & Curt.) Van Arx]	Singh et al. (1997, 2003), Holbrook and Isleib (2001), Li et al. (2012)
Leaf rust (*Puccinia arachidis* Spegazzini)	Singh et al. (1997, 2003)
Aflatoxin contamination (*Aspergillus* spp.)	Singh et al. (1997), Whitaker et al. (2004)
Stem and pod rot (*Sclerotium rolfsii*)	Singh et al. (1997), Gorbet et al. (2004)
Bacterial wilt resistance (*Ralstonia solanacearum*)	Singh et al. (1997)
Root-knot nematode (*Meloidogyne* spp.)	Holbrook and Isleib (2001)
Tomato spotted wilt virus (TSWV)	Li et al. (2012)
Groundnut rosette disease (GRD)	Subrahmanyam et al. (1998)
Kalahasty malady (*Tylenchorhynchus brevilineatus*)	Mehan et al. (1993)
Insect pests	Amin et al. (1985), Harish et al. (2012)
Abiotic stress tolerance	
Drought tolerance	Upadhyaya (2005), Jogloy et al. (2009), Balota et al. (2012)
Heat	Ntare et al. (2001)
Cold	Upadhyaya et al. (2009)
Salinity	Singh et al. (2008)
Aluminum toxicity	Boshou et al. (2000)
Nitrogen fixation tolerant to soil drying	Devi et al. (2010)
Quality parameters and others	
Nutritional quality	Wang et al. (2010), Upadhyaya et al. (2012)
Traits for confectionary uses	Dwivedi et al. (1994), Dwivedi and Nigam (2005)
Root hairs on roots and gynophores that enhance phosphorus uptake	Wissuwa and Ae (2001)
Phosphorus (P) uptake	Wissuwa and Ae (1998)

Understanding the nature of variability of the trait is important to select the breeding scheme to be employed for the improvement of the target trait. Reddy (1988) compiled the results of most of the reported genetic studies in groundnut up until 1986 in a book chapter. The results of genetic/inheritance studies on various traits in groundnut continue to appear in literature. Both, qualitative and quantitative inheritance is reported for the traits of importance. The qualitative traits, as expected, are governed by a few genes [rust (Paramasivam et al., 1990), root-knot nematode (Choi et al., 1999), rosette virus (Nigam and Bock, 1990), branching pattern (Pattanashetti et al., 2008), and high O/L ratio (Moore and Knauft, 1989)]. Both quantitative and qualitative inheritance is also reported for several traits by different workers [(resistance to LLS; Motagi et al., 2000; Dwivedi et al., 2002), fresh seed dormancy (Khalfaoui, 1991; Upadhyaya and Nigam, 1999)]. Most often, there is preponderance of additive genetic variance for quantitative traits [drought tolerance traits (Upadhyaya, 2005), and seed size (Venuprasad et al., 2011)]. Presence of dominance variance and epistasis variance due to allotetraploid nature of the crop are also reported for most of the quantitative traits, but these cannot be exploited in a self-pollinated crop such as groundnut. Inheritance of some of the traits of economic importance is not yet fully understood. In many other cases, not much is known about their inheritance pattern because of lack of effective phenotyping tools.

Studies on character association have resulted in identification of associated traits that result in maximizing gains through selection of both, target and its associated traits. In literature, several studies have demonstrated the utility of correlation analysis in groundnut based on plant and reproductive traits (Gomes and Lopes, 2005). Pod yield was reported to be positively associated with number and mass of mature pods per plant, number and mass of mature kernels per plant, shelling outturn, 100-seed mass, primary and secondary branches per plant, and harvest index. Thus, selection is exercised for these associated traits simultaneously with target trait. Sometimes character association studies have enabled the use of a robust, low-cost, and convenient surrogate trait as an alternative to a laborious trait for phenotyping and making selections. For instance, SPAD chlorophyll meter reading (SCMR) and specific leaf area (SLA) are highly correlated with water-use efficiency (WUE; Sheshashayee et al., 2006) and hence SPAD and SLA are used as the surrogate traits of WUE. At ICRISAT and several other groundnut breeding programs for drought tolerance, the trait-based approach wherein selections are based on surrogate traits of WUE, such as SPAD and SLA is adopted in combination with empirical approach to maximize the genetic gains.

The g × e has considerable influence on the progress of crop improvement and hence an important aspect for consideration. In groundnut, a majority of target traits of economic importance are polygenic and are highly influenced by environment that hinders the achievable genetic gains in breeding programs. Genetic analysis of yield revealed high influence of environment on pod yield (Zhang et al., 2011). High yielding cultivars with the least g × e interactions are normally desirable. However, when a cultivar is to be selected for a specific environment, the g × e interaction is desirable for maximizing production.

Breeding Methods

The breeding methods used for self-pollinated crops are applied in groundnut breeding. They include mass selection, pedigree, bulk, single seed descent, and back-cross methods. Introduction and mass selections played an important role in the beginning, but later, hybridization followed by selection in segregating generations following different methods was predominantly practiced in breeding improved groundnut varieties. Emasculation and pollination procedures of hybridization are cumbersome and the success rate of making crosses is generally low, particularly when carried out by inexperienced hands. Another major challenge in groundnut breeding like in many other crops is the time (8 years or more) lag between hybridizing two parents and identification of an improved breeding line for release as variety.

Segregating populations derived from crossing two parents are most common in groundnut breeding. Nevertheless, multiple crossing systems, such as the double or convergent cross, to create adequate genotypic variability before selection (Wynne and Gregory, 1981) were also used. In groundnut, pedigree and bulk-pedigree methods of breeding are most frequently used to handle segregating populations derived from hybridization. Pedigree method allows breeders to practice selection of traits with high heritability, such as plant type, pod and seed size, and shape and testa color in early generations. Selection of quantitative traits such as yield and seed composition are made in later generations. Bulk-pedigree method is a modified method of bulk aimed at improving traits with low heritability traits (Wynne and Gregory, 1981). Single seed decent method is becoming popular as this has the advantage to save space and resources (Isleib et al., 1994). Back cross breeding is used only sparingly; however with the advent of molecular markers linked to the traits of interest and quantitative trait locus (QTL) identification and mapping, back-crossing is used frequently in breeding programs. This will be discussed in more details in molecular breeding (see Groundnut Improvement Using Genomic Tools). Population improvement procedures are not commonly used in groundnut because of difficulty in making large number of crosses required for population improvement. The genetic variability available in *A. hypogaea* genetic resources is being exploited through the above breeding procedures to develop improved groundnut cultivars. But utilization of variability from wild *Arachis* species has been limited due to reproductive isolation of the cultivated species, *A. hypogaea*, an allotetraploid, from its diploid wild species due to ploidy differences and sterility. Wide hybridization has been used to harness the potential alleles from wild species and this yielding fruits with regards to resistance to disease. Despite attempts the vast potential variability of wild *Arachis* species still remains locked. Mutation breeding, using both physical and chemical mutagens has also been extensively used in groundnut breeding to induce variability that has resulted in development of several improved varieties. By 1996 in China, 15% of new groundnut varieties were bred from induced mutants and they accounted for 19.5% of the cumulative cultivated area in the country (Qui et al., 1998).

Field designs play an important role from early stages of breeding that involve testing of a large number of genotypes for selection of few elite genotypes at later stages. Thus selection of an appropriate field designs is critical for efficient and optimal utilization of resources, while ensuring identification of best possible genetic combinations in breeding programs. Multi-environment testing is adopted for development of superior genotypes adapted to a wide range of environmental conditions. Participatory plant breeding has gained significance as it allows selection of genotypes for specific adaptation rather than for broad spatial adaptation (Ceccarelli et al., 2007; Vindhiyavarman et al., 2010) and it is particularly important in developing countries as it also paves way to adoption of new improved varieties. In countries like India, Vietnam, Indonesia, and African countries, lack of adoption of new varieties has been a major bottleneck to improved groundnut yields.

Phenotyping Tools

Reliable and repeatable phenotyping remains the key to the success of any crop improvement program whether following conventional or molecular breeding approach. Several critical decisions in the process of breeding rely on results obtained from phenotyping. Phenotyping is important, first to identify a suitable source/donor for the trait, and second for selection and advancing the plants/progenies through the generations in the breeding cycles. Selecting a plant/progeny with desired combination of traits and rejecting the undesirable remains a challenging task in breeding programs as selections have to be exercised on a large number of plants/progenies with due consideration to a large number of traits simultaneously. Often, targeting multiple traits in breeding programs is limited by the challenges posed by simultaneous phenotyping for multiple traits. While in some cases it may not be feasible in others, though feasible, imposed screen for one trait may have confounding effect(s) on the other. For instance, our experience in foliar fungal disease screening nursery has shown that the rust pathogen, being obligate in nature, fails to establish and survive on leaf tissues that are already dead following leaf spot pathogen infection. As a consequence the plants infected with LLS pathogen will record low levels of rust infection and may get selected although the reason of low infection is other than resistance in this case. Another major pitfall in phenotyping is the proportion of chance escapes that get selected and hence the reliability and reproducibility are critical for a phenotyping procedure.

Phenotyping for qualitative traits, inherited by few major genes is relatively easy. Phenotyping tools are extensively used by breeders in collaboration with plant pathologists, physiologists, entomologists, biochemists, and food technologists for selection and advancement of breeding populations and evaluation of advanced breeding lines. The technique that is reproducible, robust, cost-effective, and non-destructive (for kernel oil content, etc.) is desirable for phenotyping as breeding programs in general have large number of segregating generations and advanced breeding lines to screen. Phenotyping for reaction to diseases includes screening under artificial inoculation in the laboratory, green house, and in field and under natural disease pressure at hot spot locations (Mayee and Munde, 1979; Foster et al., 1980; Nigam and Bock, 1990; Pande et al., 1994; Subrahmanyam et al., 1995; Shokes et al., 1998; Gorbet et al., 2004; Nigam et al., 2012). Either empirical approach that involves measuring the yield under imposed drought stress or salinity conditions or trait-based approach using surrogates or a combination of both are used for phenotyping abiotic stresses. Understanding of physiological mechanisms involved in abiotic stress tolerance enable use of associated physiological traits in breeding programs to enhance the efficiency of selection of desirable genotypes. Phenotyping tools for some traits such as *A. flavus* contamination in the field and reaction to soil-borne diseases need to be further developed to improve their repeatability. Similar is the case for phenotyping tools for reaction to insect pests. If surrogates for complex traits are found, they could be exploited for phenotyping of complex traits. Currently available phenotyping techniques for important target traits of groundnut breeding have been described by Janila and Nigam (2012) and are summarized in Table**2**.

Table 2. Groundnut traits for which phenotyping methods are described in the literature

Trait	Reference
Abiotic stress tolerance	
Empirical approach for drought tolerance	Rao and Nigam (2003)
Transpiration efficiency	Ratnakumar et al. (2009)
SPAD chlorophyll meter reading (SCMR), specific leaf area (SLA), Carbon isotope discrimination [and thus water-use efficiency (WUE)]	Rao et al. (2001), Sheshashayee et al. (2003)
High temperature	Craufurd et al. (2002, 2003)
Salinity	Vadez et al. (2005), Singh et al. (2008)
Aluminum toxicity	Yang and Jing (2000), Pratap et al. (2002)

Biotic stress tolerance/ resistance	
A 1–9 scale for recording reaction of leaf spot and rust diseases in the field	Subrahmanyam et al. (1995)
Detached leaf technique for leaf spots and rust	Mayee and Munde (1979), Foster et al. (1980)
Aflatoxin contamination	Mehan and McDonald (1980), Mehan (1989), Holbrook et al. (1994)
Stem and pod rot	Pande et al. (1994), Shokes et al. (1998), Gorbet et al. (2004)
Groundnut rosette virus disease	Nigam and Bock (1990), Nigam et al. (2012)
Peanut bud necrosis disease	Pensuk et al. (2002), Nigam et al. (2012)
Tomato spotted wilt virus	Mandal et al. (2001), Nigam et al. (2012)
Peanut stripe virus disease	Middleton et al. (1988), Nigam et al. (2012)
Peanut stem necrosis disease	Nigam et al. (2012)
Bacterial diseases	Li and Tan (1984), Mehan et al. (1994)
Root-knot nematode	Holbrook et al. (1983)
Aphids	Zeyong et al. (1995)
Thrips	Dwivedi et al. (1995)
Spodoptera litura, leaf miner, and jassids	Ranga Rao and Wightman (1997)
Quality traits	
Blanchability	Singh et al. (1996)
Seed coat color	Dwivedi and Nigam (2005)
Protein	Singh and Jambunathan (1980), Misra et al. (2000)
Oil content	Jambunathan et al. (1985), Misra et al. (2000)
Fatty acid	Phillips and Singleton (1981)
Iron and zinc	Sahrawat et al. (2002)
Agronomic traits and others	
Maturity duration based on the concept of cumulative thermal time or degree days (CTT or °Cd), which is both, location and season neutral	Vasudeva Rao et al. (1992)
Haulm quality parameters	Nigam and Blummel (2010)
Biological nitrogen fixation efficiency of groundnut genotypes	Wynne et al. (1980), Nambiar et al. (1982), Nigam et al. (1985), Herdina and Silsbury (1990)

Some Accomplishments of Conventional Approaches

Following the breeding procedures and using various techniques of phenotyping, several improved groundnut varieties were developed and released for cultivation world over. Since 1986, national partners in 36 countries have released 136 improved ICRISAT-bred varieties in their respective counties. Some of the varieties are released in more than one country. The national programs in India have released 189 improved groundnut varieties for cultivation in the country. The genetic gain for yield in India came from improvement in seed size, seed weight, and number of pods per plant (Ratnakumar et al., 2012). From their study, Reddy and Basu (1989) reported that improved varieties alone contributed to 30% yield increase in India since 1967. Details of some these varieties released in different countries follow.

JL 24, a popular short-duration groundnut cultivar in India, is a selection made from genotype EC 94943 introduced from Taiwan at the Oilseeds Research Station, Jalgaon, Maharashtra. It was released as JL 24 (Phule Pragati) in 1979 for cultivation in India (Patil et al., 1980). Subsequently, it was introduced to Africa and was released and commercially cultivated in several other countries (Chiyembekeza et al., 2001). It was also released in 1984/85 as Sinpadetha 2 in Myanmar and in 1992 as UPL Pn 10 in the Philippines. CG 7 (derived from USA 20 × TMV 10 cross), also known as ICGMS 42 or ICGV-SM 83708, is a high yielding Virginia bunch variety released in 1990. Recommended for cultivation in all groundnut-growing areas of Malawi, it is suitable for confectionery use and oil extraction. It is more tolerant to drought and much easier to harvest than Chalimbana, the most widely grown variety in Malawi (Subrahmanyam et al., 2000). Subsequently, it was also released in Tanzania as Mnanje in 2009, in Uganda as Serenut 1 R in 1999 and in Zambia as MGV 4 in 1990.

Following extensive screening of 5000 breeding lines and germplasm in wilt-sick plots in China and Indonesia, many genotypes with varying levels of resistance have been identified and using them bacterial wilt resistant cultivars were released in China, Indonesia, and other South East Asian countries that have offered protection against losses caused by the disease (Mehan et al., 1994; Wang et al., 2009). At ICRISAT, breeding for foliar fungal disease has resulted in development of several genotypes with significant levels of resistance to rust and LLS (Singh et al., 2003), which are either released for cultivation in several countries of Asia and Africa or have been used as parents in national breeding programs. However, the first few foliar diseases resistant releases such as Girnar 1 (derived from X 14-4-B-19-B × NC Ac 17090) in 1989 and ICG(FDRS) 10 (derived from X 14-4-B-19-B × PI 259747) in 1990 in India, in spite of their higher yields under disease epidemic conditions, did not find acceptance among the farmers due to their poor pod and kernel characteristics as a result of linkage drag.

Resistance to GRD was discovered in the late 1950s in local land races of Burkina Faso. Utilizing them, cultivars resistant to GRD, such as KH 149A, KH 241D, 69-101, RMP 12, RMP 91, and RG 1 were bred and released in the past in Africa. These cultivars are now used as sources of resistance as the land races were semi-erect and late maturing and several GRD resistant groundnut varieties have been released (Mayeux et al., 2003; Ndjeunga et al., 2003; Waliyar et al., 2007). Wide hybridization has been used to expand the gene pool by tapping alleles from wild species and several interspecific derivatives have been developed for use as donors of desirable traits or release as cultivars. One classical example of this is the release of two root-knot nematode-resistant groundnut cultivars (COAN) by the Texas Agricultural Experiment Station in 1999 after successful transfer of high level of nematode resistance into *A. hypogaea*from wild diploid species, *A. cardenasii* (Simpson and Starr, 2001). Varieties with multiple resistances were also bred following conventional methods and one such example is "Tifguard," a groundnut variety bred for resistance to both root-knot nematode and TSWV (Holbrook et al., 2008). The screening methods for nematodes were useful to identify resistant source and subsequently use the source to breed genotypes with resistance to root-knot nematodes (Simpson et al., 2003) and Kalahasti malady (Tirupati 3; Mehan et al., 1993). GPBD 4, Spanish bunch groundnut genotype resistant to rust and LLS was released for cultivation in India (Gowda et al., 2002). ICGV 86855, one of the parents of GPBD 4, is an interspecific derivative between *A. hypogaea* × *A. cardenasii* resistant to rust and LLS.

Drought resistant genotypes of groundnut were developed by selecting high pod yield under imposed soil moisture stress, referred to as empirical approach (Branch and Kvien, 1992). Several drought tolerant varieties (55-437, GC 8-35, 55-21, 55-33, SRV 1-3, and SRV

1-96 among others) were developed in West Africa (Mayeux et al., 2003). Drought tolerant groundnut variety ICGV 91114, released in India for cultivation in drought-prone district Anantapur, which has world's largest groundnut-growing area of about 800,000 ha in a district, has resulted in increase in net income by 36 and 30% reduction in year-to-year yield variability for farmers in the district (Birthal et al., 2011). In China, several new groundnut cultivars with improved productivity and some level of tolerance to aflatoxin contamination are extensively used in production in combination with integrated management approaches for aflatoxin management (Boshou et al., 2009). To increase breeding efficiency, studies on mechanisms of resistance to pre-harvest aflatoxin contamination were conducted and the most promising mechanisms identified were resistance to drought and resistance to the groundnut root-knot nematode (Holbrook et al., 2009). These approaches seem to be viable options for use in breeding in the absence of sources of resistance to aflatoxin contamination.

Varieties suitable for food uses were also released that include varieties for confectionary uses that have high seed mass, desirable seed shape, and flavor. The return of investment (ROI) was 133% for cultivation of large-seeded confectionary varieties, Asha (ICGV 86564) and Namnama (ICGV 90320) in the Philippines (PCARRD, 2009). Asha and Namnama were bred at ICRISAT. In 2002, they were introduced to Philippines and subsequently released in 2005 for Region 2 of Philippines through Bureau of Agricultural Research (BARI) and Philippines Council of Agriculture, Forestry and Natural Resources Research and Development (PCAARD). Another ICRISAT-bred variety, ICGV 00440 was released as Mallika in India. Induced mutagenesis along with recombination breeding at Bhabha Atomic Research Centre (BARC), Trombay, India succeeded in developing several large-seeded genotypes, of which TG 1, TKG 19A, Somnath (TGS 1), TPG 41, TLG 45, and TG 39 were released for cultivation to the Indian farmers (Badigannavar and Mondal, 2007). Groundnuts are bred for high oleic to linoleic ratio (O/L ratio) to improve the oil quality. Gorbet and Knauft (1997) registered the first high oleic line, SunOleic 95R, and it was followed by another variety, Hull with high O/L ratio and resistance to TSWV (Gorbet, 2007). In breeding programs targeting improvement of biological nitrogen fixation (BNF), high performing genotypes were identified at ICRISAT (ICRISAT, 1981) and North Carolina State University, Raleigh, USA (Wynne et al., 1980, 1983).

GROUNDNUT IMPROVEMENT USING GENOMIC TOOLS

The conventional breeding approaches have been successful in development and release of several improved cultivars not just for increased yield but also for resistance/tolerance to biotic and abiotic stresses thus offering protection against losses and for consumer/trader preferred traits. The approaches were discussed highlighting some successes in Section "Groundnut Improvement Using Conventional Approaches." However, the progress has been limited in improvement of some of the traits. It is possible to specifically target such traits for improvement, and also enhance the efficiency of overall breeding program through use of molecular breeding approaches. The following are the possible advantages of integration of molecular breeding approaches: (1) It provides tools to target traits of economic importance that remained poorly/not amenable to conventional breeding approaches in part due to their quantitative nature of inheritance and presence of high g × e. The identification of genomic regions, popularly called QTLs for quantitative traits is now routine, and thus it is possible to achieve genetic gains through selection for quantitative traits *via* selection of the genomic region harboring major QTLs for that trait; (2) Through molecular markers it is also possible to reduce the burden of linkage drag when the desirable traits are attempted to transfer from wild species or their derivatives into improved varieties. Systematic introgression of entire genome of a wild species into cultivated background is also possible by use of molecular markers, which is referred to as genome-wide introgression (GWI). Following this approach, it is possible to develop chromosome segment substitution lines (CSSLs) and advanced back-cross (AB)-QTL (AB-QTL) mapping populations which enable tapping of new alleles from wild species (Tanksley and Nelson, 1996). More details on these two approaches are given in Section "Emerging Genomics and Breeding Approaches"; (3) Optimum utilization of time and resources as several unwanted plants/progenies are rejected in early generations and also possibly at early growth stages of the plant based on the genotype data; (4) Phenotyping is not required in every generation as the plants/progenies can be advanced based on the genotypic data, and thus phenotyping can be postponed to later generations when non-segregating lines are derived. This results in drastic cut in the cost of phenotyping as well as increased efficiency of phenotyping when relatively less number of lines are needed to be evaluated; (5) Selection and advancing of the plants/progenies when done based on the genotyping data will eliminate the problem associated with chance failure of phenotype screen and chance escapes, which occur often with field screening. Molecular markers are unaffected by environmental fluctuations and hence, serve as reliable tags to track the target traits/genomic regions in breeding populations; (6) It is possible to improve an elite/popular cultivar for a target trait through marker-assisted back-crossing (MABC). Although back-crossing is a known method of breeding its utility in groundnut improvement remained meager till the advent of molecular technologies and there is renewed interest in this method of breeding through MABC; (7) It saves time significantly when recessive genes need to introgress/pyramid which is need of the hour in order to develop improved cultivars in shorter period of time; and (8) Gene pyramiding, i.e., targeting multiple traits is possible through molecular breeding as each of the target trait can be tracked simultaneously in the segregating populations though QTLs and/or markers.

Marker-assisted back-crossing has been the most preferred and result oriented molecular breeding approach for improving existing popular genotype for one or two traits and pyramiding of few genes/QTLs. Most importantly, MABC is now in routine use in development of near-isogenic lines (NILs) or CSSLs for genomics research. The limitation of this approach is its inability to handle multiple traits/QTLs/genes in one attempt due to requirement of large back-cross populations which becomes unmanageable (Ribaut and Hoisington, 1998; Varshney et al., 2012). Since, majority of the economically important traits are quantitative in nature and is governed by several genomics regions (genes/QTLs) distributed on plant genome and contribution of each is very small. Hence to improve such complex traits such as drought tolerance and seed yield wherein QTL analysis ends up with identification of several small-effect QTLs (Ravi et al., 2011;Gautami et al., 2012), a better molecular breeding approach called marker-assisted recurrent selection (MARS) has been proposed (Ribaut and Ragor, 2007) which can target more number of minor as well as major QTLs. MARS involves estimation of marker effects followed by two or three recombination cycles based on presence of marker alleles for small-effect QTLs (Bernardo and Charcosset, 2006; Eathington et al., 2007). Currently in groundnut, MABC is used for improvement of resistance to foliar fungal diseases, root-knot nematode and enhance oil quality for which linked markers for QTLs with high phenotypic effect were successfully identified. While, for improvement of drought tolerance in groundnut MARS was the suggested method of breeding as more than 100 main and epistatic effect QTLs were reported, it is not attempted due to lack of dense linkage maps in groundnut that limits the applicability of MARS (Ravi et al., 2011).

Markers for Target Traits

Before the linkage maps were made available, bulk segregant analysis (BSA) was used to identify the linked markers for nematode resistance (Burrow et al., 1996; Garcia et al., 1996), aphid vector of GRD (Herselman et al., 2004), and yield and yield parameters (Selvaraj et al., 2009). But the approach of identifying markers for targets traits swiftly changed with the development of linkage maps in groundnut (see Pandey et al., 2012b). QTL analysis was used for identification of QTLs for several important traits such as drought tolerance related traits (Varshney et al., 2009; Ravi et al., 2011; Gautami et al., 2012), resistance to foliar disease (Khedikar et al., 2010; Mondal et al., 2012; Sujay et al., 2012), and nutritional quality traits (Chen et al., 2010; Chu et al., 2011; Sarvamangala et al., 2011) were reported. The availability of genomic resources such as molecular markers, genetic and physical maps, expressed sequence tags (ESTs), mutant resources, and functional genomics platforms that facilitate the identification of QTLs and discovery of genes associated with tolerance/resistance to abiotic and biotic stresses and agronomic traits was reviewed (Holbrook et al., 2011; Pandey et al., 2012b). Identification of target markers to traits of economic importance has enabled integration of molecular breeding in groundnut improvement as will be discussed in the following section.

Ongoing Molecular Breeding Activities for Groundnut Improvement

Despite several possibilities to overcome the bottlenecks of conventional breeding and some demonstrated successes in other crops including groundnut, the uptake of molecular breeding approaches in groundnut improvement programs in developed countries is contrasting with that in developing countries such as India and African countries. This may in be in part due to lack of infrastructure, high cost of genotyping, and human capacities available in the developing countries. The progress made when molecular breeding was integrated with conventional procedures in development of improved groundnut varieties was quite impressive. Holbrook et al. (2011) reviewed the impact of molecular genetic research, both genomic and transgenic tools on groundnut cultivar development. In this section we dwell upon some examples of its successful deployment. The first groundnut variety developed through integrated marker-assisted selection (MAS) in the USA is NemaTAM, a root-knot nematode-resistant variety (Simpson et al., 2003). However this was susceptible to TSWV. Subsequently a cultivar named, Tifguard, bred through conventional breeding method with resistance to both TSWV and root-knot nematode was released for cultivation in the country (Holbrook et al., 2008). Following marker-assisted breeding, "Tifguard High O/L" cultivar was developed through three rounds of accelerated back-crossing to pyramid nematode resistance and the trait for high oleic:linoleic acid (high O:L) ratio in seeds (Chu et al., 2011). In this MABC program, Tifguard was used as recurrent parent.

At ICRISAT, one of the goals of breeding program is to develop high yielding varieties with resistance to rust and LLS. After identification of a major QTL ($QTL_{rust}01$) contributing up to 82.96% phenotypic variation for rust resistance, it was introgressed through MABC to improve three popular groundnut varieties (ICGV 91114, JL 24, and TAG 24) for rust resistance using GPBD 4 as a donor genotype. Several promising introgression lines with remarkable reduction in disease spread and other desirable agronomic traits have been selected for further multiplication and generation advancement (Pandey et al., 2012a). The promising genotypes with desirable yield potential and higher resistance to leaf rust could be released as improved varieties. MABC is also in progress for introgression of mutant fatty acid dehydrogenase (FAD) alleles on A and B genomes that govern high oleate trait (high oleic to linoleic acid ratio) that imparts high quality of oil benefiting both, consumers health and food processing industries (through enhanced shelf-life). The elite breeding lines with high oil content (>55%) and stable performance have been targeted for improvement of oil quality (Janila et al., 2012). In addition, gene pyramiding has also been initiated targeting one major QTL each for rust and LLS resistance after identification of major QTL for LLS resistance on AhXII that explains up to 62% of phenotypic variation from the same donor (Sujay et al., 2012).

Initiatives have been also taken at Chiba Prefectural Agriculture and Forestry Research Center in Japan, for marker-assisted introgression of mutant *FAD2* alleles into an elite cultivar "Nakateyutaka" using a breeding line "YI-0311" as donor which had an O/L ratio of 48 (Koilkonda et al., 2012). Nakateyutaka, a Virginia type is a leading variety in Japan and has normal O/L ratio. The *ahFAD2A* and *ahFAD2b* mutant alleles of the genotype YI-0311 were same as the previously reported mutational alleles found on high O/L groundnut genotypes (Jung et al., 2000) and used by Chu et al. (2011) in their MABC program. In addition to these ongoing molecular breeding activities the national programs for groundnut improvement in China and India have also initiated deployment of molecular breeding. The next decade may probably witness a good number of groundnut varieties developed though integrated molecular breeding approaches.

Tools to Tap Alleles from Wild Species

Genetic variability holds the key for the success of breeding program and groundnut has a twofold problem in this respect: first is the low genetic variability due its nature of origin, and second, the reproductive isolation from its wild diploid species due to ploidy differences and sterility. Wild *Arachis* species are known as repositories of several desirable alleles, however, wider use of wild species in breeding has been hampered by ploidy and sexual incompatibility barriers, by linkage drag, and historically, by a lack of the tools needed to conventionally confirm hybrid identities and track introgressed chromosomal segments (Bertioli et al., 2011). They remain under-utilized due to burden of linkage drag although the crossing barriers are to some extent overcome through techniques of wide hybridization. GWI and AB-QTL mapping are two important molecular marker-based approaches that enable enhanced utilization of alleles from wild species. GWI of a small genomic region from wild species while keeping the genetic background of the cultivated genotype is a good means to explore the largely untapped reservoir of useful alleles of interest in wild species. This is especially interesting in species like groundnut with narrow genetic base. This approach has been widely utilized for introgression of favorable QTLs for various traits in other crops such as tomato, rice, wheat, and barley (Fridman et al., 2004; Wang et al., 2005; Liu et al., 2006; Schmalenbach et al., 2009). AB-QTL mapping facilitates simultaneous discover of QTLs and development of elite genotypes (Tanksley and Nelson, 1996). AB-QTL mapping was used in tomato to breed an elite processing line (Tanksley et al., 1996). In groundnut, Foncéka et al. (2009) used AB-QTL approach to develop a genetic linkage map of wild genome introgression into cultivated background through utilization of synthetic amphidiploid between *A. duranensis* and *A. ipaensis*, and also derived CSSLs and AB populations. The CSSLs and AB populations facilitate characterization of different segments of genome of wild species contributing for resistance to foliar diseases and/or any other desirable trait. Once these different segments and their roles are determined, it is then possible to track them along the back-crosses using molecular markers for use in breeding programs. AB-QTL populations are also under development at ICRISAT. More recently, development of synthetic amphidiploids (Foncéka et al., 2009; Mallikarjuna et al., 2011) can facilitate better utilization of wild species in breeding programs as use of synthetic amphidiploids circumvent the crossing barriers between wild and cultivated species.

EMERGING GENOMICS AND BREEDING APPROACHES

Genome Sequence Database

Because of large genome size and amphidiploid nature the genome and heavy costs associated, it was not possible earlier to initiate genome sequencing. However due to advances in next generation sequencing and coordination of large number of partners, Peanut Genome Project

(PGP)[2] has been initiated recently with specific goals for sequencing the groundnut genome and developing genomic resources for use in groundnut improvement programs (see Pandey et al., 2012b). It is expected that draft genome sequence and extensive genomic and transcriptome information will be available soon that will enable deployment of modern genotyping approaches such as genotyping-by-sequencing (GBS). GBS is expected to make genotyping costs cheaper and faster and accessible to a broad groundnut community. As a result, for trait mapping, genome-wide association studies (GWAS) will be in routine in coming years in groundnut breeding.

Genomic Selection

Both MABC and MARS requires development of family mapping populations and identification of QTLs/marker effects before getting into the main stages of improvement programs. Further due to difficulty in handling polygenic traits and sometimes in development of good mapping populations, a better approach called "genomic selection (GS)" is fast emerging as a molecular breeding approach for crop improvement. Identification of superior lines with higher breeding value (genomic-estimated breeding values, GEBVs) in segregating breeding populations based on genome-wide marker profile data is the first step toward using this approach. To do so, a training population (TP) comprised of elite breeding lines for which multiple-season phenotyping data on agronomically important traits are available across environments is required for estimating GEBVs. Parental genotypes are then selected based on GEBVs and crosses are effected to develop candidate population (CP). In other words, CP is developed from the crosses made using the lines with best GEBVs in the TP as parents (Varshney et al., 2013). GS is now preferred over MABC and MARS for improving complex traits as GS has the advantage of selecting lines based on entire genome rather than one/few small segment of genome. It also enjoys the benefits of MABC and MARS by affecting selections based on genotype and prior of extensive phenotyping thus saving time and resources (Jannink et al., 2010). In order to exploit the recent advances in groundnut genomics to improve complex traits such as drought tolerance and seed yield, efforts have been initiated at ICRISAT to apply GS. In this direction, a TP has been developed that includes about 300 advanced breeding lines for which historical data on their performance have already been compiled.

SUMMARY

The conventional breeding approaches have largely utilized the available genetic variability in cultivated groundnut and to some extent the variability trapped in wild *Arachis* species was also used to develop improved groundnut varieties. The breeding procedures used for self-pollinated crops are employed in groundnut improvement programs along with use of phenotyping tools. Identifying and assessing the nature of variability for target traits, utilizing the sources of variability as parents in hybridization, and advancing the best possible genotypes after selection are the key steps in the groundnut breeding programs. Pedigree, bulk-pedigree, and single seed decent methods are followed to handle segregating population after hybridizing two parents. Following conventional approaches, several improved groundnut varieties with high yield and tolerance/resistance to foliar fungal diseases, bacterial wilt, root-knot nematode, virus, rosette diseases, and drought were released for cultivation. The released varieties have a wide range of maturity duration, ranging between 90 and over 150 days required for cultivation in various growing regions with varying LGP. They belonged to different market types, *viz*, Spanish, Virginia, and Valencia and meet market uses (oil and food uses) and agro-climatic requirements.

The last decade has witnessed development of molecular marker linkage maps in groundnut that was followed by identification of markers and QTLs for target traits. This paved way for deployment of molecular tools in breeding program for efficient utilization of time and resources and improved efficiency of breeding. As a consequence, the extensive breeding programs are becoming intensive with the use of molecular breeding tools. Marker technologies offer approaches (GWI and AB-QTL mapping) that enable tapping of new alleles from wild *Arachis* species that remain under-utilized. Targeting multiple traits (gene pyramiding) is another important possibility of molecular breeding approaches. Groundnut breeding programs in USA, China, India, and Japan have already embarked the new technology to complement the ongoing breeding programs. "NemaTAM" is the first cultivar developed through molecular breeding for resistance to root-knot nematode. This was followed by "Tifguard High O/L" that has high O/L ratio and multiple resistances. At ICRISAT, MABC is underway to develop cultivars with rust resistance that are now in advance generations and also pyramid resistance to rust and LLS. MABC for improvement of oil quality is underway at USA, Japan, and India. The uptake of molecular breeding tools in groundnut breeding programs in developed countries is contrasting with developing countries like India and African countries, which is in part due to inadequate infrastructure, high genotyping costs, and inadequate human capacities in the later. On the other hand, more efficient genomic tools are under development, and this is expected to happen more rapidly once the draft genome sequence is available.

ACKNOWLEDGMENTS

The authors profoundly acknowledge the financial assistance received from CGIAR Generation Challenge Programme (GCP) and Bill & Melinda Gates Foundation (BMGF) through Tropical Legumes-I (GCP, BMGF) and Tropical Legumes-II (BMGF) projects for supporting the development and deployment of genomic tools and genetic resources as well as breeding and extension activities, and its national partners in South-Asia and Sub-Saharan Africa.

REFERENCES

1. Amin P. W., Singh K. N., Dwivedi S. L., Rao V. R. (1985). Sources of resistance to the jassids (*Empoasca kerri* Pruthi), thrips (*Frankliniella schultzei* (Trybom)) and termites (*Odontotermes* sp.) in groundnut (*Arachis hypogaea* L.). *Peanut Sci.* 12 58–60
2. Badigannavar A. M., Mondal S. (2007). Mutation experiments and recent accomplishments in Trombay groundnuts. *IANCAS Bull.* 6 308–318
3. Balota M. T., Isleib G., Tallury S. (2012). Variability for drought related traits of Virginia-type peanut cultivars and advanced breeding lines. *Crop Sci.* 52 2702–2713

4. Bernardo R., Charcosset A. (2006). Usefulness of gene information in marker-assisted recurrent selection: a simulation appraisal. *Crop Sci.* 46 614–621
5. Bertioli D. J., Seijo G., Freitas F. O. (2011). An overview of peanut and its wild relatives. *Plant Genet. Resour.* 9 134–149
6. Birthal P. S., Nigam S. N., Narayanan A. V., Kareem K. A. (2011). *An Economic Assessment of the Potential Benefits of Breeding for Drought Tolerance in Crops: A Case of Groundnut in India*.Research Bulletin no. 25. Patancheru: International Crops Research Institute for the Semi-Arid Tropics
7. Birthal P. S., Rao P. P., Nigam S. N., Bantilan C. S., Bhagavatulu S. (2010). *Groundnut and Soybean Economies in Asia: Facts, Trends and Outlook*. Patancheru: International Crops Research Institute for the Semi-Arid Tropics
8. Boshou L., Rong Z., Yong L., Dong L. (2000). Evaluation of tolerance to aluminum toxicity in high-yielding groundnut genotypes. *Chin. J. Oil Crop Res.* 22 38–42
9. Boshou L., Zhuang W., Tang R., Zhang X., Shan S., Jiang H., et al. (2009). Peanut aflatoxin and genomics research in China: progress and perspectives. *Peanut Sci.* 36 21–28
10. Branch W. D., Kvien C. K. (1992). Cytoplasmically inherited albinism in peanut seedlings. *J. Hered.* 83 455–457
11. Burrow M. D., Simpson C. E., Paterson A. H., Starr J. L. (1996). Identification of peanut (*Arachis hypogaea* L.) RAPD markers diagnostic of root-knot nematode (*Meloidogyne arenaria* (Neal) Chitwood) resistance. *Mol. Breed.* 2 369–379
12. Ceccarelli S., Grando S., Baum M. (2007). Participatory plant breeding in water-limited environments. *Exp. Agric.* 43 411–435
13. Chen Z., Wang M. L., Barkley N. A., Pittman R. N. (2010). A simple allele-specific PCR assay for detecting FAD2 alleles in both A and B genomes of the cultivated peanut for high-oleate trait selection. *Plant Mol. Biol. Rep.* 28 542–548
14. Chiyembekeza A. J., Subrahmanyam P, vander Merwe P. J. A. (2001). *Rural Prosperity is Nation's Economic Stability: A Partnership Approach to Attain Sustainable Production of Groundnut and Pigeonpea in Smallholder Agriculture for Quality Diet, Household Food Security, and Poverty Alleviation in*

Malawi. Quarterly Report (January to March 2001). Lilongwe: USAID, ICRISAT in partnership with The Ministry of Agriculture and Irrigation, Malawi

15. Choi K., Burow M. D., Church G., Burow G., Paterson A. H., Simpson C. E., et al. (1999).Genetics and mechanism of resistance to *Meloidogyne arenaria* in peanut germplasm. *J. Nematol.*31 283–290
16. Chu Y., Wu C. L., Holbrook C. C., Tillman B. L., Person G., Ozias-Akins P. (2011). Marker-assisted selection to pyramid nematode resistance and the high oleic trait in peanut. *Plant Genome* 4110–117
17. Craufurd P. Q., Vara Prasad P. V., Summerfield R. J. (2002). Dry matter production and rate of change of harvest index at high temperature in peanut. *Crop Sci.* 42 146–151 [
18. Craufurd P. Q., Vara Prasad P. V., Kakani V. G., Wheeler T. R., Nigam S. N. (2003). Heat tolerance in groundnut. *Field Crops Res.* 80 63–77
19. Devi M. J., Sinclair T. R., Vadez V. (2010). Genotypic variability among peanut (*Arachis hypogea*L.) in sensitivity of nitrogen fixation to soil drying. *Plant Soil* 330 139–148
20. Dwivedi S. L., Crouch J. H., Nigam S. N., Ferguson M. E., Paterson A. H. (2003). Molecular breeding of groundnut for enhanced productivity of security of food security in the semi-arid tropics: opportunities and challenges. *Adv. Agron.* 80 221
21. Dwivedi S. L., Nagabhushanam G. V. S., Nigam S. N., Raghunath K., Jambunathan R. (1994).Germplasm enhancement for seed-quality traits in groundnut. *Int. Arachis Newsl.* 14 14–15
22. Dwivedi S. L., Nigam S. N. (2005). Confectionery groundnuts: issues and opportunities to promote export and food uses in India. *J. Oilseeds Res.* 22 1–4
23. Dwivedi S. L., Nigam S. N., Reddy D. V. R., Reddy A. S, Ranga Rao G. V. (1995). "Progress in breeding groundnut varieties resistant to peanut bud necrosis virus and its vector," in *Recent Studies on Peanut Bud Necrosis Disease: Proceedings of a Meeting, March 20, 1995*, eds Buiel A. A. M., Parlevliet J. E., Lenne J. M., editors. (Patancheru: International Crops Research Institute for the Semi-Arid Tropics/Wageningen: Department of Plant Breeding, Wageningen Agricultural University;) 35–40

24. Dwivedi S. L., Pande S., Rao J. N., Nigam S. N. (2002). Components of resistance to late leaf spot and rust among interspecific derivatives and their significance in a foliar disease resistance breeding in groundnut (*Arachis hypogaea* L.). *Euphytica* 125 81–88

25. Dwivedi S. L., Upadhyaya H. D., Blair W. M., Bartioli D. J., Nielen S., Ortiz R. O. (2008).Enhancing crop gene pools with beneficial traits using wild relatives. *Plant Breed. Rev.* 30 179–230

26. Eathington S. R., Crosbie T. M., Edwards M. D., Reiter R. S., Bull J. K. (2007). Molecular markers in a commercial breeding program. *Crop Sci.* 47 154–163

27. FAOSTAT (2010). Available at: http://faostat.fao.org/ (accessed on October 15, 2012).

28. Fonceka D., Hodo-Abalo T., Rivallan R., Faye I., Sall M. N., Ndoye O., et al. (2009). Genetic mapping of wild introgressions into cultivated peanut: a way toward enlarging the genetic basis of a recent allotetraploid. *BMC Plant Biol.* 9:103 10.1186/1471-2229-9-103 [[[Cross Ref

29. Foster D. J., Wynne J. C., Beute M. K. (1980). Evaluation of detached leaf culture for screening peanuts for leaf spot resistance. *Peanut Sci.* 7 98–100

30. Fridman E., Carrari F., Liu Y. S., Fernie A. R., Zamir D. (2004). Zooming in on a quantitative trait for tomato yield using interspecific introgressions. *Science* 305 1786–1789 [

31. Garcia G. M., Stalker H. T., Shoroeder E., Kochert G. (1996). Identification of RAPD, SCAR, and RFLP markers tightly linked to nematode resistance genes introgressed from *Arachis cardenasii* into*Arachis hypogaea*. *Genome* 39 836–845 [

32. Gautami B., Pandey M. K., Vadez V., Nigam S. N., Ratnakumar P., Krishnamurthy L., et al. (2012). QTL analysis and consensus genetic map for drought tolerance traits based on three RIL populations of cultivated groundnut (*Arachis hypogaea* L.). *Mol. Breed.* 32 757–772[[

33. Gomes R. L. F, Lopes A. C. A. (2005). Correlations and path analysis in peanut. *Crop Breed. Appl. Biotechnol.* 5 105–112

34. Gorbet D. W. (2007). Registration of 'Hull' peanut. *J. Plant Regist.* 1 125–126

35. Gorbet D. W., Knauft D. A. (1997). Registration of 'SunOleic 95R' peanut. *Crop Sci.* 37 1392
36. Gorbet D. W., Kucharek T. A., Shokes F. M., Brenneman T. B. (2004). Field evaluations of peanut germplasm for resistance to stem rot caused by *Sclerotium rolfsii. Peanut Sci.* 31 91–95
37. Gowda M. V. C., Motagi B. N., Naidu G. K., Diddimani S. N., Sheshagiri R. (2002). GPBD 4: a Spanish bunch groundnut genotype resistant to rust and late leaf spot. *Int. Arachis Newsl.* 22 29–32
38. Gregory W. C., Krapovickas A., Gregory M. P. (1980). "Structure, variation, evolution, and classification in *Arachis*," in *Advances in Legume Science*, Vol. 1 eds Summerfield R. J., Bunting A. H., editors. (Kew: Royal Botanic Gardens;) 469–481
39. Harish G., Rathnakumar A. L., Kumar N., Ajay B. C., Holajjer P., Savaliya S. D., et al. (2012).Comparative response of groundnut genotypes to bruchid beetle, *Caryedon serratus* Oliver in storage. *Ann. Plant Prot. Sci.* 20 140–144
40. Herdina J. A., Silsbury J. H. (1990). Estimating nitrogenase activity of faba bean (*Vicia faba* L.)by acetylene reduction (AR) assay. *Aust. J. Plant Physiol.* 17 489–502
41. Herselman L., Thwaites R., Kimmins F. M., Courtois B., van der Merwe P. J. A., Seal S. E. (2004).Identification and mapping of AFLP markers linked to peanut (*Arachis hypogaea* L.) resistance to the aphid vector of groundnut rosette disease. *Theor. Appl. Genet.* 109 1426–1433 [
42. Holbrook C. C., Anderson W. F., Pittman R. W. (1993). Selection of a core collection from the U.S. germplasm collection of peanut. *Crop Sci.* 33 859–861
43. Holbrook C. C., Jr., Dong W. (2005). Development and evaluation of a mini-core collection for the U.S. peanut germplasm collection. *Crop Sci.* 45 1540–1544
44. Holbrook C. C., Guo B. Z., Wilson D. M., Timper P. (2009). The U.S. breeding program to develop peanut with drought tolerance and reduced aflatoxin contamination. *Peanut Sci.* 36 50–53
45. Holbrook C. C., Isleib T. G. (2001). Geographical distribution of genetic diversity in *Arachis hypogaea*. *Peanut Sci.* 28 80–84

46. Holbrook C. C., Knauft D. A., Dickson D. W. (1983). A technique for screening peanut for resistance to *Meloidogyne arenaria*. *Plant Dis.* 67 957–958

47. Holbrook C. C., Matheron M. E., Wilson D. W., Anderson W. F., Will M. E., Norden A. J. (1994).Development of a large-scale field screening system for resistance to pre-harvest aflatoxin contamination. *Peanut Sci.* 21 20–22

48. Holbrook C. C., Ozias-Akins P., Chu Y., Guo B. (2011). Impact of molecular genetic research on peanut cultivar development. *Agronomy* 1 3–17

49. Holbrook C. C., Timper P., Culbreath A. K., Kvien C. K. (2008). Registration of 'Tifguard' peanut.*J. Plant Regist.* 2 92–94

50. ICRISAT. (1981). *ICRISAT Annual Report 1981*. Patancheru: International Crops Research Institute for the Semi-Arid Tropics

51. Isleib T. G., Wynne J. C., Nigam S. N. (1994). "Groundnut breeding," in *The Groundnut Crop: A Scientific Basis for Improvement*, ed. Smartt J., editor. (London: Chapman & Hall;) 552–623

52. Jambunathan R. (1991). "Groundnut quality characteristics," in *Uses of Tropical Grain Legumes: Proceedings of a Consultants Meeting, March 27–30, 1989* (Patancheru: ICRISAT;), 267–275

53. Jambunathan R., Raju M. S., Barde S. P. (1985). Analysis of oil content of groundnuts by nuclear magnetic resonance spectrometry. *J. Sci. Food Agric.* 36 162–166

54. Janila P., Nigam S. N. (2012). *Phenotyping for Groundnut (Arachis hypogaea L.) Improvement*. Berlin: Springer (in press)

55. Janila P., Pandey M. K., Nagesh M. P., Manohar S. S., Roorkiwal M., Sriswathi M., et al. (2012)."Marker-assisted introgression of high oleate trait to improve oil quality in elite genotypes of groundnut (*Arachis hypogaea* L.)," in *VI International Conference on Legume Genetics and Genomics (ICLGG) Programme and Abstract Book (2–7 October 2012, Hyderabad, India).*

56. Jannink J. L., Lorenz A. J., Iwata H. (2010). Genomic selection in plant breeding: from theory to practice. *Brief. Funct. Genomics* 9 166–177 [

57. Jiang H. F., Ren X. P., Liao B. S., Huang J. Q., Lei Y., Chen B. Y., et al. (2008). Peanut core collection established in china

and compared with ICRISAT mini core collection. *Acta Agron. Sin.*34 25–30

58. Jiang H. F., Ren X. P., Zhang X. J., Huang J. Q., Lei Y., Yan L. Y., et al. (2010). Comparison of genetic diversity based on SSR markers between peanut mini core collections from China and ICRISAT. *Acta Agron. Sin.* 36 1084–1091
59. Jogloy P. S., Vorasoot N., Akkasaeng C., Kesmala T., Patanothai A. (2009). Variability in yield responses of peanut (*Arachis hypogaea* L.) genotypes under early season drought. *Asian J. Plant Sci.* 8 254–264
60. Johansen C, Nageswara Rao R. C. (1996). "Maximizing groundnut yields," in *Achieving High Groundnut Yields: Proceedings of International Workshop, August 25–29, 1995, Laixi City, Shandong,* ed. Renard C., Gowda C. L. L., Nigam S. N., Johansen C., editors. (Patancheru: ICRISAT;) 117–127
61. Jung S., Swift D., Sengoku E., Patel M., Teulé F., Powell G., et al. (2000). The high oleate trait in the cultivated peanut (*Arachis hypogaea* L.). I. Isolation and characterization of two genes encoding microsomal oleoyl-PC desaturases. *Mol. Gen. Genet.* 263 796–805 [
62. Kalyani G., Reddy A. S., Kumar P. L., Rao R. D. V. J. P., Aruna R., Waliyar F., et al. (2007).Sources of resistance to tobacco streak virus in wild *Arachis* (Fabaceae: Papilionoideae) germplasm. *Plant Dis.* 91 1585–1590
63. Khalfaoui J. L. B. (1991). Inheritance of seed dormancy in a cross between two Spanish peanut cultivars. *Peanut Sci.* 18 65–67
64. Khedikar Y. P., Gowda M. V. C., Sarvamangala C., Patgar K. V., Upadhyaya H. D., Varshney R. K. (2010). A QTL study on late leaf spot and rust revealed one major QTL for molecular breeding for rust resistance in groundnut (*Arachis hypogaea* L.). *Theor. Appl. Genet.* 121 971–984[[
65. Kochert G., Stalker H. T., Gimenes M., Galgaro L., Lopes C. R., Moore K. (1996). RFLP and cytogenetic evidence on the origin and evolution of allotetraploid domesticated peanut *Arachis hypogaea* (Leguminosae). *Am. J. Bot.* 83 1282–1291
66. Koilkonda P., Kuwata C., Fukami M., Shirasawa K., Aoki K., Tabata S., et al. (2012). "Marker assisted backcrossing selection for high

O/L ratio in cultivated peanut," in *VI International Conference on Legume Genetics and Genomics (ICLGG) Programme and Abstract Book (2–7 October 2012, Hyderabad, India)*.

67. Krapovickas A., Gregory W. C. (1994). Taxonomía del género Arachis (Leguminosae).*Bonplandia* 8 1–186
68. Li W. R., Tan Y. J. (1984). Inoculation techniques for groundnut bacterial wilt. *Oil Crop Chin.* 277–81
69. Li Y., Culbreath A. K., Chen C. Y., Knapp S. J., Holbrook C. C., Guo B. (2012). Variability in field response of peanut genotypes from the U.S. and China to tomato spotted wilt virus and leaf spots. *Peanut Sci.* 39 30–37
70. Liu S. B., Zhou R. H., Dong Y. C., Li P., Jia J. Z. (2006). Development utilization of introgression lines using a synthetic wheat as donor. *Theor. Appl. Genet.* 112 1360–1373 [
71. Lynch R. E. (1990). Resistance in peanut to major arthropod pests. *Fla. Entomol.* 73 422–445
72. Mallikarjuna N., Senthilvel S., Hoisington D. A. (2011). Development of new sources of tetraploid*Arachis* to broaden the genetic base of cultivated groundnut (*Arachis hypogaea* L.). *Genet. Resour. Crop Evol.* 58 889–907
73. Mandal B., Pappu H. R., Culbreath A. K. (2001). Factors affecting mechanical transmission of tomato spotted wilt virus to peanut (*Arachis hypogaea*). *Plant Dis.* 85 1259–1263
74. Mayee C. D., Munde P. N. (1979). A modified detached leaf technique for laboratory development of groundnut rust. *Indian Phytopathol.* 32 467
75. Mayeux A. H., Waliyar F., Ntare B. R. (2003). *Groundnut Varieties Recommended by the Groundnut Germplasm Project (GGP) for West and Central Africa (In En., Fr.)*. Patancheru: International Crops Research Institute for the Semi-Arid Tropics, 80 p
76. Mehan V. K. (1989). "Screening groundnuts for resistance to seed invasion by *Aspergillus flavus*and to aflatoxin production" in *Aflatoxin Contamination of Groundnut: Proceeding of International Work-shop* ed. McDonald D., Mehan V. K., Hall S. D., editors. (Patancheru: International Crops Research Institute for the Semi-Arid Tropics;) 323–334

77. Mehan V. K., Liao B. S., Tan Y. J., Robinson-Smith A., McDonald D., Hayward A. C. (1994). *Bacterial Wilt of Groundnut*. Information Bulletin no.35. Patancheru: International Crops Research Institute for the Semi-Arid Tropics

78. Mehan V. K., McDonald M. (1980). *Screening for Resistance to Aspergillus flavus Invasion and Aflatoxin Production in Groundnuts*. ICRISAT, Groundnut Improvement Program Occasional Paper no. 2. Patancheru: ICRISAT (limited distribution)

79. Mehan V. K., Reddy D. D. R., McDonald D. (1993). Resistance in groundnut genotypes to Kalahasti malady caused by the stunt nematode, *Tylenchorhynchus brevilineatus*. *Int. J. Pest Manage* 39 201–203

80. Middleton K. J., Saleh N., Horn N., Reddy D. V. R. (1988). "Production of peanut stripe virus-free seed and screening for resistance to peanut stripe virus," in *Summary Proceedings of the First Meeting to Coordinate Research on Peanut Stripe Virus Disease of Groundnut* (Malang: Malang Research Institute for Food Crops/Patancheru: International Crops Research Institute for the Semi-Arid Tropics;), 21–22

81. Misra J. B., Ghosh P. K., Dayal D., Mathur R. S. (2000). Agronomic, nutritional and physical characteristics of some Indian groundnut cultivars. *Indian J. Agric. Sci.* 70 741–746

82. Mondal S., Badigannavar A. M, D'Souza S. F. (2012). Development of genic molecular markers linked to a rust resistance gene in cultivated groundnut (*Arachis hypogaea* L.). *Euphytica* 188 163–173

83. Moore K. M., Knauft D. A. (1989). The inheritance of high oleic acid in peanut. *J. Hered.* 80 252–253

84. Moretzsohn M. C., Gouvea E. G., Inglis P. W., Leal-Bertioli S. C. M., Valls J. F. M., Bertioli D. J. (2012). A study of the relationships of cultivated peanut (*Arachis hypogaea*) and its most closely related wild species using intron sequences and microsatellite markers. *Ann. Bot.* 111 113–126[[

85. Motagi B. N., Gowda M. V. C., Naidu G. K. (2000). Inheritance of late leafspot resistance in groundnut mutants. *Indian J. Genet. Plant Breed.* 60 347–352

86. Nambiar P. T. C., Dart P. J., Nigam S. N., Gibbons R. W. (1982). "Genetic manipulation of nodulation in groundnut," in *Biological Nitrogen Fixation Technology for Tropical Agriculture*, eds Graham P. H., Harris S. C., editors. (Cali: Centro Internacional de Agricultura Tropical;) 49–56

87. Nath U. K., Alam M. S. (2002). Genetic variability, heritability and genetic advance of yield and related traits of groundnut (*Arachis hypogaea* L.). *J. Biol. Sci.* 2 762–764

88. Nautiyal P. C., Rajgopal K., Zala P. V., Pujari D. S., Basu M. S., Dhadhal B. A., et al. (2008).Evaluation of wild *Arachis* species for abiotic stress tolerance: thermal stress and leaf water relations. *Euphytica* 159 43–57

89. Ndjeunga J., Ntare B. R., Waliyar F., Kodio O., Traore T. (2003). Assessing diffusion of modern groundnut varieties in Mali. *Int. Arachis Newsl.* 23 33–35

90. Nigam S. N. (2000). Some strategic issues in breeding for high and stable yield in groundnut in India. *J. Oilseeds Res.* 17 1–10

91. Nigam S. N., Blummel M. (2010). Cultivar-dependent variation in food-feed-traits in groundnut (*Arachis hypogaea* L.). *Anim. Nutr. Feed Technol.* 10S 39–48

92. Nigam S. N., Bock K. R. (1990). Inheritance of resistance to groundnut rosette virus in groundnut (*Arachis hypogaea* L.). *Ann. Appl. Biol.* 117 558–560

93. Nigam S. N., Dwivedi S. L., Gibbons R. W. (1991). Groundnut breeding: constraints, achievements and future possibilities. *Plant Breed. Abstr.* 61 1127–1136

94. Nigam S. N., Dwivedi S. L., Nambiar P. T. C., Gibbons R. W., Dart P. J. (1985). Combining ability analysis of N2-fixation and related traits in peanut. *Peanut Sci.* 12 55–57

95. Nigam S. N., Prasada Rao R. D. V. J., Bhatnagar-Mathur P., Sharma K. K. (2012). "Genetic management of virus diseases in peanut," in *Plant Breeding Reviews* Vol. 36 ed. Janick J., editor. (New York: John Wiley & Sons, Inc./Wiley-Blackwell;) 293–356

96. Nigam S. N., Rao V. R., Gibbons R. W. (1983). Utilization of natural hybrids in the improvement of groundnuts (*Arachis hypogaea*). *Exp. Agric.* 19 355–359

97. Ntare B. R., Williams J. H., Dougbedji F. (2001). Evaluation of groundnut genotypes for heat tolerance under field conditions in a Sahelian environment using a simple physiological model for yield. *J. Agric. Sci.* 136 81–88

98. Pande S., Narayana Rao J., Reddy M. V., McDonald D. (1994). Development of a greenhouse screening technique for stem rot resistance in groundnut. *Int. Arachis Newsl.* 14 23–24

99. Pandey M. K., Janila P., Nigam S. N., Gowda M. V. C., Manohar S. S., Sriswathi M., et al. (2012a). "Disease screening and yield assessment of introgression lines developed through marker-assisted backcrossing for rust resistance in cultivated groundnut (*Arachis hypogaea* L.)," in *VI International Conference on Legume Genetics and Genomics (ICLGG) Programme and Abstract Book (2–7 October 2012, Hyderabad).*

100. Pandey M. K., Monyo E., Ozias-Akins P., Liang X., Guimarães P., Nigam S. N., et al. (2012b).Advances in *Arachis* genomics for peanut improvement. *Biotechnol. Adv.* 30 639–651 [

101. Paramasivam K., Jayasekhar M., Rajsekharan R., Veerbadhiran P. (1990). Inheritance of rust resistance in groundnut (*A. hypogaea* L.). *Madras Agric. J.* 77 50–52

102. Patil G. D., Desale S. C., Patil P. S., Patil S. S. (1980). 'Phule-Pragati' a high yielding early bunch groundnut for Maharashtra. *J. Maharashtra. Agric. Univ.* 5 47–52

103. Pattanashetti S. K., Gowda M. V. C, Girija. (2008). Inheritance of morphological traits and pod features in groundnut (*Arachis hypogaea* L.). *Indian J.Genet. Plant Breed.* 68 157–162

104. Pattee H. E., Isleib T. G., Giesbrecht F. G., McFeeters R. F. (2000). Relationship of sweet, bitter and roasted peanuts sensory attributes with carbohydrate components in peanuts. *J. Agric. Food Chem.* 48 757–763 [

105. PCARRD. (2009). *Asha and Namnama Peanut Confectionery Varieties*. Information Bulletin no.254. Manila: Department of Science and Technology, Philippines Council of Agriculture, Forestry, Natural Resources Research and Development (PCARRD)

106. Pensuk V., Daengpluang N., Wongkaew S., Jogloy S., Patanothai A. (2002). Evaluation of screening procedures to identify peanut resistance to peanut bud necrosis virus (PBNV). *Peanut Sci.*29 47–51

107. Phillips R. J., Singleton B. (1981). The determination of specific free fatty acids in peanut oil by gas chromatography. *J. Am. Oil Chem. Soc.* 55 225–227

108. Pratap S., Ramarao G., Reddy K. B. (2002). Response of groundnut genotypes to aluminum toxicity in solution culture. *Indian J. Plant Physiol.* 7 396–400

109. Qui Q. S., Li Z. C., Shen Y. F., Wang C. T., Miao H. R. (1998). Groundnut breeding through mutation techniques in China. *Int. Arachis Newsl.* 18 17

110. Ramanathan T. (2004). "Groundnut," in *Applied Genetics of Oilseed Crops*, ed. Ramanathan T. (New Delhi: Daya Publishing House;) 15–73

111. Ranga Rao G.V., Wightman J. A. (1997). "Techniques for screening groundnut genotypes for resistance to insect pests," in *Spodoptera litura in India: Proceedings of the National Scientist Forum on spodoptera litura, April 2–4, 1996* (Patancheru: ICRISAT/London: Overseas Development Administration;), 67–75

112. Rao N., Reddy L., Bramel P. (2003). Potential of wild species for genetic enhancement of some semi-arid food crops. *Genet. Resour. Crop Evol.* 50 707–721

113. Rao R. C. N., Nigam S. N. (2003). "Genetic options for drought management in groundnut," in*Management of Agricultural Drought – Agronomic and Genetic Options*, eds Saxena N. P., editor. (New Delhi: Publisher Oxford & IBH Publishing Co. Pvt. Ltd.) 123–141

114. Rao R. C. N., Talwar H. S., Wright G. C. (2001). Rapid assessment of specific leaf area and leaf nitrogen in peanut (*Arachis hypogaea* L.) using a chlorophyll meter. *J. Agron. Crop Sci.* 189 175–182

115. Ratnakumar A. L., Hariprasanna K., Lalwani H. B. (2012). Genetic improvement in Spanish type groundnut, *Arachis hypogaea* L. varieties in India over the years. *J. Oilseeds Res.* 27 1–7

116. Ratnakumar P., Vadez V., Nigam S. N., Krishnamurthy L. (2009). Assessment of transpiration efficiency in peanut (*Arachis hypogaea* L.) under drought using a lysimeter system. *Plant Biol.* 11, 124–130 [

117. Ravi K., Vadez V., Isobe S., Mir R. R., Guo Y., Nigam S. N. (2011). Identification of several small main-effect QTLs and a large number of epistatic QTLs for drought tolerance related traits in groundnut (*Arachis hypogaea* L.). *Theor. Appl. Genet.* 122 1119–1132 [[

118. Reddy P. S. (1988). "Genetics, breeding and varieties," in *Groundnut*, ed. Reddy P. S., editor. (New Delhi: Indian Council of Agricultural Research;) 200–317

119. Reddy P. S., Basu M. S. (1989). *Groundnut Varieties of India*. Technical Bulletin. Junagadh: National Research Center for Groundnut, ICAR; 15

120. Ribaut J. M., Hoisington D. (1998). Marker-assisted selection: new tools and strategies. *Trends Plant Sci.* 3 236–239

121. Ribaut J. M., Ragor M. (2007). Marker-assisted selection to improve drought adaptation in maize: the backcross approach, perspectives, limitations, and alternatives. *J. Exp. Bot.* 58 351–360[

122. Sahrawat K. L., Ravi Kumar G., Rao J. K. (2002). Evaluation of triacid and dry ashing procedures for determining potassium, calcium, magnesium, iron, zinc, manganese, and copper in plant materials. *Commun. Soil Sci. Plant Anal.* 33 95–102

123. Sarvamangala C., Gowda M. V. C., Varshney R. K. (2011). Identification of quantitative trait loci for protein content, oil content and oil quality for groundnut (*Arachis hypogaea* L.). *Field Crops Res.* 122 49–59

124. Schmalenbach I., Leon J., Pillen K. (2009). Identification and verification of QTLs for agronomic traits using wild barley introgression lines. *Theor. Appl. Genet* 118 483–497 [

125. Selvaraj M. G., Narayana M., Schubert A. M., Ayers J. L., Baring M. R., Burow M. D. (2009).Identification of QTLs for pod and kernel traits in cultivated peanut by bulked segregant analysis.*Electron. J. Biotechnol.* 12. 10.2225/vol12-issue2-fulltext-13 [Cross Ref

126. Sharma S. B., McDonald D. (1990). Global status of nematode problems of groundnut, pigeonpea, chickpea, sorghum and pearl millet and suggestion for future work. *Crop Prot.* 9 453–458

127. Sheshashayee M. S., Bindumadhavi H., Rachaputi N. R., Prasad T. G., Udayakumar M., Wright G. C., et al. (2006). Leaf chlorophyll

concentration relates to transcription efficiency in peanut. *Ann. Appl. Biol.* 148 7–15

128. Sheshashayee M. S., Bindhumadhava H., Shankar A. G., Prasad T. G., Udayakumar M. (2003).Breeding strategies to exploit water use efficiency for crop improvement. *J. Plant Biol.* 30 253–268
129. Shokes F. M., Weber Z., Gorbet D. W., Pudelko H. A., Taczanowski M. (1998). Evaluation of peanut genotypes for resistance to southern stem rot using an agar disc technique. *Peanut Sci.* 2512–17
130. Simpson C. E., Starr J. L. (2001). Registration of 'COAN' peanut. *Crop Sci.* 41 918
131. Simpson C. E., Starr J. L., Church G. T., Burrow M. D., Paterson A. H. (2003). Registration of NemaTAM peanut. *Crop Sci.* 43 1561
132. Simpson C. E., Starr J. L., Nelson S. C., Woodard K. E., Smith O. D. (1993). Registration of TxAG-6 and TxAG-7 peanut germplasm. *Crop Sci.* 33 1418
133. Singh A. K., Dwivedi S. L., Pande S., Moss J. P., Nigam S. N., Sastry D. C. (2003). Registration of rust and late leaf spot resistance peanut germplasm lines. *Crop Sci.* 43 440–441
134. Singh A. K., Mehan V. K., Nigam S. N. (1997). *Sources of Resistance to Groundnut Fungal and Bacterial Diseases: An Update and Appraisal.*Information Bulletin no. 50. Patancheru: ICRISAT
135. Singh A. K., Simpson C. E. (1994). "Biosystematics and genetic resources," in *The Groundnut Crop: A Scientific Basis for Improvement*, ed. Smartt J., editor. (London: Chapman & Hall;), 96–137
136. Singh A. L., Hariprassana K., Solanki R. M. (2008). Screening and selection of groundnut genotypes for tolerance of soil salinity. *Aust. J. Crop Sci.* 1 69–77
137. Singh F., Diwakar B. (1993). *Nutritive Value and Uses of Pigeonpea and Groundnut. Manual.*Patancheru: International Crops Research Institute for the Semi-Arid Tropics
138. Singh U., Jambunathan R. (1980). Evaluation of rapid method for estimation of protein in chickpea (*Cicer arietinum* L.). *J. Sci. Food Agric.* 31 247–254

139. Singh U., Sridhar R., Dwivedi S. L., Nigam S. N., Jambunathan R. (1996). Evaluation of blanching quality in groundnut (*Arachis hypogaea* L.). *J. Food Sci. Technol.* 33 211–214

140. Subrahmanyam P., Hildebrand G. L., Naidu R. A., Reddy L. J., Singh A. K. (1998). Sources of resistance to groundnut rosette disease in global groundnut germplasm. *Ann. Appl. Biol.* 132 473–485

141. Subrahmanyam P., McDonald D., Waliyar F., Reddy L. J., Nigam S. N., Gibbons R. W., et al. (1995). *Screening Methods and Sources of Resistance to Rust and Late Leaf Spot of Groundnut.* Information Bulletin no.47. Patancheru: International Crops Research Institute for the Semi-Arid Tropics

142. Subrahmanyam P., Van der Merwe P. J. A., Chiyembekeza A. J., Ngulube S., Freeman H. A. (2000). Groundnut variety CG 7: a boost to Malawian Agriculture. *Int. Arachis Newsletter* 20 33–35

143. Sujay V., Gowda M. V. C., Pandey M. K., Bhat R. S., Khedikar Y. P., Nadaf H. L., et al. (2012).Quantitative trait locus analysis and construction of consensus genetic map for foliar disease resistance based on two recombinant inbred line populations in cultivated groundnut (*Arachis hypogaea* L.). *Mol. Breed.* 30 773–788 [[

144. Swamy B. P. M., Upadhyaya H. D, Goudar P. V. K. (2006). Characterization of Asian core-collection of groundnut for morphological traits. *Indian J. Crop Sci.* 1 129–134

145. Tanksley S. D., Grandillo S., Fulton T. M., Zamir D., Eshed Y. Petiard. V., et al. (1996). Advanced backcross QTL analysis in a cross between an elite processing line of tomato and its wild relative *L. pimpinellifolium*. *Theor. Appl. Genet.* 92 213–224 [

146. Tanksley S. D., Nelson J. C. (1996). Advanced backcross QTL analysis: a method for the simultaneous discovery and transfer of valuable QTLs from unadapted germplasm into elite breeding lines. *Theor. Appl. Genet.* 92 191–203 [

147. UNICEF. (2007). Available at: http://www.unicef.org/infobycountry/niger_39675.html (accessed on October 24, 2012)

148. Upadhyaya H. D. (2005). Variability for drought resistance related traits in the mini core collection of peanut. *Crop Sci.* 45 1432–1440

149. Upadhyaya H. D., Bramel P. J., Ortiz R., Singh S. (2002). Developing a mini.core of peanut for utilization of genetic resources. *Crop Sci.* 42 2150–2156
150. Upadhyaya H. D., Dronavalli N., Singh S., Dwivedi S. L. (2012). Variability and stability for kernel iron and zinc contents in ICRISAT mini core collection of Peanut. *Crop Sci.* 52 2628–2637
151. Upadhyaya H. D., Dwivedi S. L., Nadaf H. L., Singh S. (2011). Phenotypic diversity and identification of wild Arachis accessions with useful agronomic and nutritional traits. *Euphytica* 82103–115
152. Upadhyaya H. D., Nigam S. N. (1999). Inheritance of fresh seed dormancy in peanut. *Crop Sci.* 3998–101
153. Upadhyaya H. D., Ortiz R., Bramel P. J, Singh S. (2003). Development of a groundnut core collection using taxonomical, geographical and morphological descriptors. *Genet. Resour. Crop Evol.* 50 139–148
154. Upadhyaya H. D., Reddy L. J., Dwivedi S. L., Gowda C. L. L., Singh S. (2009). Phenotypic diversity in cold-tolerant peanut (*Arachis hypogaea* L.) germplasm. *Euphytica* 165 279–291
155. Upadhyaya H. D., Reddy L. J., Gowda C. L. L., Singh S. (2006). Identification of diverse groundnut germplasm: sources of early maturity in a core collection. *Field Crops Res.* 97 261–271
156. Vadez V., Srivastava N., Krishnamurthy L., Aruna R., Nigam S. N. (2005). Standardization of a protocol to screen for salinity tolerance in groundnut. *Int. Arachis Newsl.* 25 42–47
157. Varshney R. K., Kudupa H., Roorkiwal M., Thudi M., Pandey M. K., Saxena R. K., et al. (2012).Advances in genomics research and molecular breeding applications in SAT legume crops by using next generation sequencing and high-throughput genotyping technologies. *J. Biosci.* 37 811–820[
158. Varshney R. K., Mahendar T., Aruna R., Nigam S. N., Neelima K., Vadez V., et al. (2009). High level of natural variation in a groundnut (*A. hypogaea* L.) germplasm collection assayed by selected informative SSR markers. *Plant Breed.* 128 486–494
159. Varshney R. K., Mohan S. M., Gaur P. M., Gangarao N. V. P. R., Pandey M. K., Bohra A., et al. (2013). Achievements and prospects

of genomics-assisted breeding in three legume crops of the semi-arid tropics. *Biotechnol. Adv.* 1–55 [

160. Vasudeva Rao M. J., Nigam S. N, Huda A. K. S. (1992). The thermal time concept as a selection criterion for earliness in peanut. *Peanut Sci.* 19 7–10
161. Venuprasad R., Aruna R., Nigam S. N. (2011). Inheritance of traits associated with seed size in groundnut (*Arachis hypogaea* L.). *Euphytica* 181 169–177
162. Vindhiyavarman P., Manivannan N., Nigam SN., Muralidharan V. (2010). Farmers' participatory varietal selection in groundnut: a case study from Tamilnadu, India. *Electron. J. Plant Breed.* 1878–881
163. Waliyar F., Kumar P. L., Ntare B. R., Monyo E., Nigam S. N., Reddy A. S., et al. (2007). *A Century of Research on Groundnut Rosette Disease and its Management.*Information Bulletin no.75. Technical Report. Patancheru: International Crops Research Institute for the Semi-Arid Tropics
164. Wang C. T., Wang X. Z., Tang Y. Y., Chen D. X., Cui F. G., Zhang J. C., et al. (2009). Field screening of groundnut genotypes for resistance to bacterial wilt in Shandong province in China. *J. SAT Agric. Res.* 7 1–5
165. Wang M. L., Chen C. Y., Davis J., Stalker T., Pittman R. N., Holbrook C. C., Jr., et al. (2010)."Assessment of oil content and fatty acid composition variability in the U.S. peanut minicore," in*ASA-CSSA-SSSA Annual Meeting Abstracts 295*, 5
166. Wang M. L., Chen C. Y., Pinnow D. L., Barkley N. A., Pittman R. N., Lamb M., Pederson G. A. (2012). Seed dormancy variability in the U.S. peanut mini-core collection. *Res. J. Seed Sci.* 5 84–95
167. Wang Y. M., Dong Z. Y., Zhang Z. J., Lin X. Y., Shen Y., Zhou D., et al. (2005). Extensive de novo genomic variation in rice induced by introgression from wild rice (*Zizania latifolia* Griseb.).*Genetics* 170 1945–1956 [[
168. Whitaker T. B., Dorner J. W., Giesbrecht F. G., Slate A. B. (2004). Variability among aflatoxin test results on runner peanuts harvested from small field plots. *Peanut Sci.* 31 59–63
169. Wightman J. A., Amin P. W. (1988). Groundnut pests and their control in semiarid tropics. *Trop. Pest Manag.* 34 218–226

170. Wissuwa M., Ae N. (1998). Genotypic variation for phosphorus uptake from hardly soluble iron-phosphate in groundnut (*Arachis hypogaea* L.). *Plant Soil* 206 163–171

171. Wissuwa M., Ae N. (2001). Genotypic differences in the presence of hairs on roots and gynophores of peanuts (*Arachis hypogaea* L.) and their significance for phosphorus uptake. *J. Exp. Bot.* 521703–1710 [

172. Wynne J. C., Elkan G. H., Schneeweis T. J. (1980). "Increasing nitrogen fixation of groundnut by strain and host selection," in *Proceedings of the International Workshop on Groundnuts, October 13–17, 1980* (Patancheru: International Crops Research Institute for the Semi-Arid Tropics;), 95–109

173. Wynne J. C., Elkan G. H., Isleib T. J., Schneeweis T. J. (1983). Effect to host plant, *Rhizobium*strain and host × strain interaction on symbiotic variability in peanut. *Peanut Sci.* 10 110–114

174. Wynne J. C., Gregory W. C. (1981). Peanut breeding. *Adv. Agron.* 34 39–72

175. Yang Q., Jing H. B. (2000). The effect of aluminum on N, P and Ca uptake in peanut varieties.*Chin. J. Oil Crops Res.* 22 68–73

176. Zeyong X., Wightman J. A., Reddy D. V. R., Zongyi Z., Kunrong C., Jinxing C., et al. (1995).Evaluation of an aphid-resistant groundnut genotype (EC 36892) in China. *Int. Arachis Newsl.* 1549–51

177. Zhang X., Han S., Tang F., Xu J., Liu H., Yan M., et al. (2011). Genetic analysis of yield in peanut (*Arachis hypogaea* L.) using mixed model of major gene plus polygene. *Afr. J. Biotechnol.* 107126–7130

Chapter 8

Kazusa Marker DataBase: A Database for Genomics, Genetics, and Molecular breeding in Plants

Kenta Shirasawa, Sachiko Isobe, Satoshi Tabata, and Hideki Hirakawa*

Department of Plant Genome Research, Kazusa DNA Research Institute, Kisarazu, Chiba 292-0818, Japan

ABSTRACT

In order to provide useful genomic information for agronomical plants, we have established a database, the Kazusa Marker DataBase (http://marker.kazusa.or.jp). This database includes information on DNA markers, e.g., SSR and SNP markers, genetic linkage maps, and physical maps, that were developed at the Kazusa DNA Research Institute. Keyword searches for the markers, sequence data used for marker development, and experimental conditions are also available through this database. Currently, 10 plant species have been targeted: tomato (*Solanum lycopersicum*), pepper (*Capsicum annuum*), strawberry (*Fragaria* × *ananassa*), radish (*Raphanus sativus*), *Lotus japonicus*, soybean (*Glycine max*), peanut (*Arachis hypogaea*), red clover (*Trifolium pratense*), white clover (*Trifolium repens*), and eucalyptus (*Eucalyptus camaldulensis*). In addition, the number of plant species registered in this database will be increased as our research progresses. The Kazusa Marker DataBase will be a useful tool for both basic and applied sciences, such as genomics, genetics, and molecular breeding in crops.

INTRODUCTION

DNA polymorphisms and mutations, some of which confer phenotypic variations, can be detected by DNA marker analyses. The first DNA marker, the restriction fragment length polymorphism (RFLP), has been used for linkage analysis to determine the genomic positions responsible for Huntington's disease in humans (Gusella *et al.* 1983). The RFLP marker technologies were then applied to analysis of plant genetics, particularly tomato and maize (Helentjaris *et al.* 1986), to construct genetic maps, which are essential tools for positional cloning and quantitative trait loci (QTL) analysis of genes of interest. Then, several types of DNA markers, e.g., random amplified polymorphic DNA (RAPD), amplified fragment length polymorphism (AFLP), simple sequence repeat (SSR) or microsatellite, and single nucleotide polymorphism (SNP), were made available through advances in the technologies for DNA analysis (Phillips and Vasil 2001).

DNA markers are used not only in basic sciences but also in applied studies (Kumar 1999). For example, DNA markers linking to desirable loci are used in the selection of elite lines from breeding populations, a process known as marker-assisted selection (MAS). F_1 hybrids are used in the production of commercial varieties in several cereal and vegetable crops, because the F_1 hybrids sometimes exhibit hybrid vigor and heterosis. Therefore, DNA markers can also be applied for purity testing to investigate the heterogeneity of F_1 hybrids, which is a combination of different alleles from the parental lines. In addition, in the management of genetic resources and quality control of food products, DNA markers have been employed for identification of species, cultivars, and varieties.

Whole-genome sequencing in plants was first achieved in *Arabidopsis thaliana* (The Arabidopsis Genome Initiative 2000), followed by rice (International Rice Genome Sequencing Project 2005). Since those initial reports, the genomes of more than 50 plants have been sequenced (Michael and Jackson 2013). In addition, massive transcriptome analysis has been performed in several plants by using the next-generation sequencers (NGSs) (Hamilton and Buell 2012). This has enabled the development of large numbers of DNA markers for several plants in a relatively short time. Numerous databases have been made available with the genome and DNA marker information for these plant species, e.g. TAIR for *Arabidopsis thaliana*(http://www.arabidopsis.org) and IRGSP for *Oryza sativa* (http://rgp.dna.affrc.go.jp/IRGSP/). In addition, there are also databases integrating the information for several plant species, such as the Gramene database for cereal plant species (http://www.gramene.org), the Sol Genomics Network (SGN) for the Solanaceae family (http://solgenomics.net), and VegMarks for seven species of vegetables analyzed in the NARO Institute of Vegetable and Tea Sciences (http://vegmarks.nivot.affrc.go.jp). Our group has also worked on DNA marker development, genetic linkage map, and QTL analysis to promote breeding programs for crops whose molecular genetic information has lagged behind those of the model species, but which are nonetheless important for food production, animal feedings, and industrial materials. In this paper, we introduce the Kazusa Marker DataBase (http://marker.kazusa.or.jp), which provides information on DNA markers, genetic linkage and physical maps developed in the Kazusa DNA Research Institute, for crops and other plant species to enrich the available molecular genetic information on agronomical plants.

CONTENTS OF THE KAZUSA MARKER DATABASE

The Kazusa Marker DataBase was constructed using Red Hat Enterprise Linux Server release 5.6 as the computer operating system. The MySQL (http://dev.mysql.com) system, which is a relational database management system, was employed for management of the database contents. Most of the contents of this database were written using html text format and Ruby on Rails (RoR), the open-source web framework (http://rubyonrails.org).

Currently, the database includes mainly SSR markers for 10 species: tomato (*Solanum lycopersicum*), pepper (*Capsicum annuum*), strawberry (*Fragaria* × *ananassa*), radish (*Raphanus sativus*), *Lotus japonicus*, soybean (*Glycine max*), peanut (*Arachis hypogaea*), red clover (*Trifolium pratense*), white clover (*Trifolium repens*), and eucalyptus (*Eucalyptus camaldulensis*) (Table 1). SSR markers have advantages over other marker systems because of their multi-allelic detection, co-dominant inheritance, high-transferability across species, and flexibility with various laboratory systems. The SSR markers can be classified into two categories, genome-SSR and expressed sequence tag (EST)-SSR. The genome-SSR markers are developed from random genome sequences from, for example, SSR-enrichment genomic libraries, while EST-SSR markers are from sequences of cDNAs. In the subsequent *in silico* analyses, SSR motifs were identified from the sequence data with the SSRIT (Temnykh *et al.* 2001), MISA (Thiel *et al.* 2003), and/or SciRoKo (Kofler*et al.* 2007) programs, and PCR primers were designed on the flanking sequences of the SSR motifs using the Primer3 program (Rozen and Skaletsky 2000). The identified SSR motifs, repeat numbers of the motif, PCR primers, and expected amplicon sizes are also available from the database. In addition, allele information and/or gel images are also available for the SSR markers, which were tested in molecular experiments. This database also includes information on SNPs, which are the most abundant source of variation in the genome for both intragenic and intergenic regions. While the SNP data are limited to tomato at present, the number of crops with SNP data is expected to increase through the use of NGSs. The details for other markers are described in the next section for each crop.

Table 1. Types and numbers of DNA markers registered in the Kazusa Marker DataBase

Binomial nomenclature	Common name	Marker type	Abbreviation of marker name	No. of markers	Reference
Solanum lycopersicum	Tomato	Genome-SSR	TGS	13,501	Shirasawa *et al.* 2010a
		EST-SSR	TES	7,599	Shirasawa *et al.* 2010a
		Genome-SNP		1,473,798	Shirasawa *et al.* 2013b
		EST-SNP		5,607	Shirasawa *et al.* 2010b
		Intron-SNP	TEI	169	Shirasawa *et al.* 2010a
Capsicum annuum	Capsicum	EST-SSR	CaES	5,751	Shirasawa *et al.* 2013c
Fragaria × ananassa	Strawberry	EST-SSR	FAES	603	Isobe *et al.* 2013
		SSR dervied from *F. vesca*	FVES	3,746	Isobe *et al.* 2013
		Transcriptome-SSR	FATS	125	Isobe *et al.* 2013
Raphanus sativus	Radish	EST-SSR	RRS	3,811	Shirasawa *et al.* 2011
Lotus japonicus		dCAPS	BM, TM	82	Sato *et al.* 2001
		SSR	BM, TM	1,073	Sato *et al.* 2001
Glycine max	Soybean	EST-SSR	GMES	7,020	Hisano *et al.* 2007
Arachis hypogaea	Peanut	EST-SSR	AHS	3,187	Koilkonda *et al.* 2012
		Genome-SSR	AHGS	6,706	Shirasawa *et al.* 2012b
		Transposable element	AhTE	1,039	Shirasawa *et al.* 2012a, 2012b
Trifolium pratense	Red clover	SSR	RCS, TPSSR	7,262	Isobe *et al.* 2009
Trifolium repens	White clover	EST-SSR	WCS	1,993	Isobe *et al.* 2012
Eucalyptus camaldulensis	Eucaly	Genome-SSR	EcGAS	4,656	Hirakawa *et al.* 2011
		EST-SSR	EcES	1,028	Hirakawa *et al.* 2011

Table 2. Genetic linkage maps registered in the Kazusa Marker DataBase

Binomial nomenclature	Common name	Map name	No. of linkage groups	No. of mapped loci	Total length (cM)	Mean marker density (cM/loci)	Reference
Solanum lycopersicum	Tomato	Tomato-EXPEN2000	12	2,116	1,503	0.7	Shirasawa *et al.* 2010a
		AMF2	12	990	1,468	1.5	Shirasawa *et al.* 2010b
		MMF2	13	637	1,230	2.0	Shirasawa *et al.* 2010b
Fragaria × ananassa	Strawberry	Integrated map	28	1,861	1,967	1.1	Isobe *et al.* 2013
Raphanus sativus	Radish	GHRI	9	843	1,129	1.4	Shirasawa *et al.* 2011
Lotus japonicus		MG-20 × B-129	6	1,155	421	0.4	Hayashi *et al.* 2001
Glycine max	Soybean	Map1	20	693	2,688	4.0	Hisano *et al.* 2007
Arachis hypogaea	Peanut	SKF2	21	1,114	2,226	2.0	Shirasawa *et al.* 2012b
		NYF2	19	326	1,333	4.3	Shirasawa *et al.* 2012b
		AF5	10	597	544	0.9	Shirasawa *et al.* 2013a
		BF5	10	798	461	0.6	Shirasawa *et al.* 2013a
		TF6	20	1,469	1,442	1.0	Shirasawa *et al.* 2013a
		Integrated map	20	3,693	2,651	0.7	Shirasawa *et al.* 2013a
Trifolium pratense	Red clover	HR × R130	7	1,714	834	0.5	Isobe *et al.* 2009

Genetic linkage maps, which were constructed with the DNA markers for seven crops, i.e., tomato, peanut, radish, soybean, red clover, *L. japonicus*, and strawberry, are also registered in this database (Table 2). The information provided on the genetic linkage maps includes the mapped positions of the various markers. For tomato, pepper, and radish, the DNA markers were mapped on the reference sequences themselves or their relatives.

USAGE INSTRUCTION FOR THE KAZUSA MARKER DATABASE

The top page of the Kazusa Marker DataBase represents crops registered in this database. Users can click either "Images", "Scientific names" in the table, or icons below the table to access pages of each crop, which include "Keyword Search", "Marker List", "Reference list", "Linkage Map", "Physical Map", and "Markers on the Genome" depending on crops (see below section). Thorough the "Keyword Search", marker names, sequence names used for marker designing, and descriptions in comment boxes can be searched. The "Marker List" contained marker types, e.g., genome-SSR and EST-SSR, as described in the below section and Table 1. By selecting the marker type, lists of the markers comprised of "Marker Name", "Marker Type", and primer sequences are available. Then, clicking marker names enables users to obtain all information on the markers, e.g., sequence name corresponding the markers with a hyperlink to public DNA sequence databases, map positions (if available), PCR fragment size estimated from the sequence, experimental conditions such as methods on PCR and detections, SSR motif and the repeated number (if SSRs), gel images (if available), and reference articles. The "Markers on the Genome" is available to presume physical genome positions of the markers, if the genome sequences of the crop itself or its relatives are released. Although bulk data download is not supported in the current version, it is available upon request to pj.ro.asuzak@bdrekram as well as to us.

PLANT SPECIES REGISTERED IN THE KAZUSA MARKER DATABASE

Tomato

Tomato (*S. lycopersicum*), an important fruit crop throughout the world and a model for fresh fruit research, is an autogamous diploid species (2n = 2x = 24) with a genome of 900 Mb, and its sequences have been published by a multinational project consortium (The Tomato Genome Consortium 2012). The database contains information on DNA markers, as well as genetic linkage and physical maps. The DNA markers include both SSR and SNP markers. The SSR markers, 7,599 EST-SSR (TES) and 13,501 genome-SSR (TGS) designed from EST and BAC-end sequences available from a public database (http://solgenomics.net), respectively, were developed to construct an interspecific high-density genetic linkage map (Shirasawa *et al.* 2010a), on which totals of 648 TES and 634 TGS were mapped. In addition, 674 EST-derived intronic polymorphism markers (TEI) were developed and 151 TEI markers were mapped (Shirasawa *et al.* 2010a). The SNP markers were also developed from EST and genome sequence data. Each of the EST-derived SNPs was developed from the alignment data of EST sequences derived from at least two tomato lines. From this analysis, 5,607 SNPs were identified in 2,634 contigs, and 793 were mapped on the two genetic linkage maps based on intraspecific crossings (Shirasawa *et al.* 2010b). On the other hand, the genome-SNPs were discovered by the re-sequencing strategy (Shirasawa *et al.* 2013b), in which sequence reads for six tomato lines by the ABI-5500xl SOLiD (DRA accession numbers: DRA001017 to DRA001022) were mapped onto the tomato reference genome, SL2.40 (The Tomato Genome Consortium 2012). A total of 1,473,798 genome-SNPs were identified and 1,536 SNPs were employed for genotyping of 663 tomato accessions stocked in gene banks (Shirasawa *et al.* 2013b). The SNPs are searchable by accession names, the genome positions, or seven categories depending on their effects on gene function, e.g., intergenic, intron, splice site, untranscribed regions, synonymous, missense, and nonsense SNPs. Furthermore, 170,173 of the SNP-derived cleaved amplified polymorphic sequence (CAPS) markers, in which 19 restriction enzymes are employed, are also available. The positions of the DNA markers developed in this study were identified on the published tomato genome (The Tomato Genome Consortium 2012). Then, the DNA markers and the predicted genes in the tomato genome were ordered in parallel based on the physical positions of the reference genome, from which the users can obtain information on the DNA markers and the predicted genes. This tool is very useful to search for DNA markers or loci in genes of interest.

In addition, we have established a portal website for tomato genomics, KatomicsDB (Shirasawa and Hirakawa 2013: http://www.kazusa.or.jp/tomato/), because our group provides not only information on the DNA markers and genetic maps as described above but also inferences of SNP effects on gene functions and sequence data of gene-rich regions in the tomato genome. The KatomicsDB contains links to the marker database described above, a functional SNP database (Hirakawa *et al.* 2013a:http://plant1.kazusa.or.jp/tomato/), and a database for genome sequences of selected BAC clone mixtures in gene-rich regions (http://www.kazusa.or.jp/tomato_sbm/).

Pepper

Capsicum spp., including *C. annuum*, *C. baccatum*, *C. chinense*, *C. frutescens*, and *C. pubescens*, belong to the Solanaceae family, and are widely cultivated for use as vegetables and spices. Like tomato, all these species are autogamous diploids (2n = 2x = 24), while the size of their genomes (~ 3.3 Gb: Moscone *et al.*2003) is more than three times larger than that of the other members of Solanaceae, e.g., tomato, potato, and eggplant. The pepper marker database includes mainly *Capsicum* EST-SSR (CaES) information. A total of 5,751 CaES markers were designed from the 118,060 EST sequences for *Capsicum annuum* obtained from a public DNA database, GenBank (http://www.ncbi.nlm.nih.gov). The CaES markers were mapped on the tomato genome by *in silico* analysis based on sequence similarity search, which is recognized as a model for the Solanaceae, because the genome sequence of pepper has not been reported (at the time of writing). However, the genome structures between tomato and pepper are conserved and exhibit a macrosynteny relationship (Wu *et al.* 2010). Therefore, the positions of the pepper DNA markers and genes on the pepper genome can be speculated by mapping them onto the tomato genome (Shirasawa *et al.* 2013c). As a result, the positions of 2,245 of the CaES markers were identified on the tomato genome. Among the 2,245 markers, 96 CaES markers were subjected to genotyping analysis of 192 *Capsicum* accessions, which have been stocked at the Kihara Institute for Biological Research of the Yokohama City University, Japan, to reveal their genetic diversity. The polymorphism information content (PIC) values and allele sizes for the 192 accessions are also available from this database. As additional markers, SNPs detected in the *matK* and *rbcL*genes coded in the chloroplast genome, which are known as "barcode" sequences for the identification of species (CBOL Plant Working Group 2009), are also available for the 192 accessions (Accession numbers:AB721552 to AB721935).

Strawberry

Strawberry (*F.* × *ananassa*) is a popular fruit cultivated throughout the world, and possesses a complex genome structure due to its octoploid nature (2n = 8x = 56) and its allogamous reproductive system. The genome size of strawberry is estimated to be 692 Mb (Hirakawa *et al.* 2013b). A wild diploid species, *F. vesca,* is one of the probable ancestral species, and 240 Mb of its genome has been sequenced (Shulaev *et al.*2011). This database includes three types of SSR markers and an integrated map. The SSR markers were designed from EST sequences of not only *F.* × *ananassa* but also *F. vesca,* because a larger number of EST sequences for *F. vesca* were available from public DNA databases than for *F.* × *ananassa*. A total of 3,746 SSRs derived from ESTs for *F. vesca,* 603 SSRs derived from ESTs for *F.* × *ananassa,* and 125 SSRs derived from transcriptomes for *F.* × *ananassa* markers were developed and subjected to map constructions (Isobe *et al.* 2013). Three genetic linkage maps were established using three mapping populations, and integrated into a consensus map consisting of 28 linkage groups with 1,856 loci, the number of which corresponded to the haploid chromosome number of *F.* × *ananassa*. In addition to the map constructions, the SSR markers were employed for the genetic diversity analysis of 129 strawberry cultivars. A total of 45 SSR markers were determined to be sufficient to distinguish 129 *F.* × *ananassa* lines except for four lines.

Radish

Radish (*R. sativus*), or Japanese daikon, is an allogamous species due to its self-incompatibility system, and has a diploid genome (2n = 2x = 18), sizing of 526 Mb (Arumuganathan and Earle 1991). The radish is a vegetable crop and a member of the Brassicaceae, to which the genera *Arabidopsis* and *Brassica* also belong, but the genomic research on radish has not been progressed as far as for members of the Brassicaceae. The daikon marker database includes mainly EST-SSR markers. A total of 3,800 radish EST-SSR markers (RSS) were developed from 26,606 EST sequences (Accession numbers: FY428055 to FY454660) (Shirasawa *et al.* 2011). Genetic linkage maps of 630 RSS markers and 213 previously reported markers were obtained from this database. Subsequent comparative analysis of the *Raphanus* map with the *Arabidopsis* and *B. rapa*genomes (The Arabidopsis Genome Initiative 2000, The *Brassica rapa* Genome Sequencing Project Consortium 2011) revealed the genomic synteny between the two species. Therefore, the radish DNA markers were *in silico* mapped on the genomes of *Arabidopsis* and *B. rapa* to speculate on the positions of the radish DNA markers and genes on the radish genome. This analysis revealed the positions of 3,234 and 3,730 SSR markers on the *Arabidopsis* and *B. rapa* genomes, respectively.

Lotus Japonicus

L. japonicus is not a crop but is recognized as a model for legume crops and symbiosis research because of its rapid life cycle, fixed genotypes due to autogamous reproduction, simple and compact genome (2n = 2x = 12, 472 Mb), and easy transformability (Handberg and Stougaard 1992). The marker database for *Lotus japonicus* consists of DNA markers and their linkage map. A total of 1,073 SSR and 82 derived CAPS (dCAPS) markers were developed by comparative analysis of the genome sequences from two *L. japonicus*strains, Miyakojima MG-20 and Gifu B-129 (Sato *et al.* 2001). A genetic linkage map of the SSR and dCAPS markers, which consisted of six linkage groups covering 1,155 cM in total, were generated by using an F_2 mapping population derived from a cross between the MG-20 and B-129 (Hayashi *et al.* 2001). By using this linkage map as a reference, the genome sequences of the

MG-20 were anchored to the chromosomes of *L. japonicus* (Sato *et al.* 2008: http://www.kazusa.or.jp/lotus/).

Soybean

Because soybean (*G. max*) is a major crop that is important for oil and protein production, its genome (2n = 2x = 40, genome size of 1.1 Gb) was sequenced in spite of the complexity of its paleopolyploidy (Schmutz *et al.* 2010). The database includes EST-SSR markers and a genetic linkage map. A total of 6,920 EST-SSR markers were developed from 63,676 publicly available non-redundant soybean ESTs from public databases (Dana-Farber Center Institute; http://compbio.dfci.harvard.edu/tgi/). Among them, 693 SSR marker loci were combined with 242 RFLP, genome-SSR, and phenotypic markers. The resultant maps consisting 20 linkage groups covered 2,700.3 cM in a total length (Hisano *et al.* 2007). The transferability of the 686 mapped markers was investigated for 24 *Glycine* accessions. The EST-SSR markers were *in silico* mapped on the genome sequences to identify the positions of the EST-SSR markers on the soybean genome.

Peanut

Peanut (*A. hypogaea*), or groundnut, is an autogamous allotetraploid (2n = 4x = 40) legume species with a genome of approximately 2.8 Gb (Arumuganathan and Earle 1991). It is used for food and oil production, and its probable ancestral species have been identified as *A. duranensis* and *A. ipaënsis*. The database consists of information on DNA markers and genetic linkage maps. As the DNA markers, a total of 6,706 genome-SSR (AHGS), 3,187 EST-SSR (AHS), and 1,039 transposon insertion length polymorphism markers (AhTE) have been developed from the sequence data collected from SSR-enriched genomic libraries (accession numbers: DH964238 to DH968256) (Shirasawa *et al.* 2012b), cDNA libraries (accession numbers: FS960760 to FS988327) (Koilkonda *et al.* 2012), and transposon-enriched genomic libraries (accession numbers: DE998420 to DE998923 and DH968257 to DH968767) (Shirasawa *et al.* 2012a,2012b), respectively. The genome- and EST-SSR and transposon markers were subjected to constructions of five genetic linkage maps in *Arachis*, SKF2 and NYF2 for cultivated peanut (*A. hypogaea*), AF5 and BF6 for wild diploid relatives (*A. duranensis*, *A. stenosperma*, *A. ipaënsis*, and *A. duranensis*), and TF6 or *A. hypogaea* and an artificial amphidiploid (*A. ipaënsis* × *A. duranensis*)^{4x} (Shirasawa *et al.* 2013a). In addition, the five genetic linkage maps were integrated with 11 published maps from other research groups under collaborations between Japan, Brazil, India, France, the US, and China (Shirasawa *et al.* 2013a). The EST-SSR markers were employed for the genetic diversity analysis of peanut accessions, including 17 Japanese, 4 American, 2 Indian, and 1 Chinese cultivated lines as well as 6 wild relatives (Koilkonda *et al.* 2012).

Red Clover

Red clover (*T. pratense*) is an allogamous diploid legume (2n = 2x = 14, genome size of 468 Mb:Arumuganathan and Earle 1991) that is cultivated as a forage crop. The database for red clover consists of information on the DNA markers, RFLP and SSR markers, and genetic linkage map. The RFLP markers were developed to construct a genetic linkage map in red clover. The resultant map contains 157 RFLP markers and covers 535.7 cM in total (Isobe *et al.* 2003). Subsequently, 7,262 SSR markers were developed from 26,356 EST sequences (Accession numbers: BB902456 to BB928811), and employed to generate a genetic linkage map consisting of 1,434 marker loci covering 868.7 cM in total (Sato *et al.* 2005). Finally, additional new linkage maps together with the developed genetic linkage maps were integrated into the consensus map with 1,804 marker loci covering 836.6 cM in total (Isobe *et al.* 2009). The resultant genetic linkage map, i.e., HR × R130, is available from this database.

White Clover

White clover (*T. repens*) is an allogamous allotetraploid legume (2n = 4x = 32, genome size of 999 Mb:Arumuganathan and Earle 1991) widely cultivated as a forage crop. We generated the white clover linkage maps using SSR markers in order to conduct comparative genomics analyses among legume species (Isobe *et al.* 2012). In this database, a total of 1,993 primers are available for the EST-derived SSR markers.

A total of 15,214 EST sequences used for primer construction are also available through the accession numbersFY454661 to FY469874.

Eucalyptus

E. camaldurensis is a diploid species (2n = 2x = 22, genome size of 650 Mb) that is used in the pulp industry. Therefore, the genome sequencing of *E. camaldulensis* and development of markers have been performed to survey the genetic information and accelerate the process of molecular breeding (Hirakawa *et al.* 2011: http://www.kazusa.or.jp/eucaly/). The eucalyptus marker database consists of information on 4,656 genome- and 1,028 EST-SSR markers, which were developed from the sequence data of the transcriptome and genome of *E. camaldulensis*, respectively. The SSR markers were employed for the genetic diversity analysis of six *Eucalyptus* species, i.e., *E. camaldulensis, E. dunnii, E. globules, E. grandis, E. nitens* and *E. urophylla*. The PIC values based on this analysis are also available from this database.

MARKER DENSITIES IN THE GENETIC AND PHYSICAL MAPS

Marker density, or mean distances between any neighboring marker intervals, would be important information for gene mapping studies of a map-based cloning strategy or genome-wide association studies (GWAS), and for MAS in breeding programs. On the one hand, as for the linkage maps registered in the database, the marker densities between any neighboring loci were varied from 0.4 cM in *L. japonicus* to 4.3 cM in peanut, and 1.5 cM in average over the 14 genetic maps of the six species (Table 2). On the other hand, as for the physical maps among the ten species, tomato had the highest dense marker loci due to the massive SNP data from the re-sequencing analysis (Table 3). The markers were estimated to locate in every 600 bp interval in the tomato genome. In the remaining nine species, mean physical intervals of any neighboring two markers were ranging from 64 kb in red clover to 573 kb in *Capsicum* (Table 3). While availability of whole genome sequence data at present were limited to tomato (The Tomato Genome Consortium

2012), strawberry (Hirakawa *et al.* 2013b), *L. japonicus* (Sato *et al.* 2008), soybean (Schmutz *et al.* 2010), and eucalyptus (Hirakawa *et al.* 2011) among the plant species registered in the Kazusa Marker DataBase at the time of writing, genome sequences for the other species would be determined in near future with the cooperation of the NGSs as summarized in Genomes OnLine Database (http://www.genomesonline.org). The whole genome sequence data identify the physical genome positions of the DNA markers, and provide useful information for the gene mapping studies as well as MAS.

Table 3. The estimated genome sizes and marker densities of the plant species in the Kazusa Marker DataBase

Binomial nomenclature	Common name	Total no. of markers	The estimated genome size (Mb)	Mean marker density (kb/loci)
Solanum lycopersicum	Tomato	1,500,674	900	0.6
Capsicum annuum	Capsicum	5,751	3,300	573.8
Fragaria × ananassa	Strawberry	4,474	692	154.7
Raphanus sativus	Radish	3,811	526	138.0
Lotus japonicus		1,155	472	408.7
Glycine max	Soybean	7,020	1,115	158.8
Arachis hypogaea	Peanut	10,932	2,813	257.3
Trifolium pratense	Red clover	7,262	468	64.4
Trifolium repens	White clover	1,993	999	501.3
Eucalyptus camaldulensis	Eucaly	5,684	650	114.4

FUTURE DIRECTIONS

Until now, the Kazusa Marker Database includes information on DNA markers, genetic linkage maps, and physical maps for 10 plant species comprised of mainly crops. Because our research groups have been working on more than 25 plant species, the contents of this database will increase when we publish papers on each project. Databases for the DNA markers have been globally established in each crop species, institutes, and countries, a situation which is considered to be undesirable for users. To overcome the problem, an integrated database of the plant genome-related information, i.e., PGDBj (http://pgdbj.jp), has been established (Asamizu *et al.* 2013), which includes parts of the marker and map information registered in the Kazusa Marker DataBase. In addition, we are planning to provide graphical views of the marker positions on genome sequences or linkage maps by using GBrowse (Stein *et al.* 2002) or CMap (Fang *et al.* 2003) from the Kazusa Marker DataBase. The user-friendly interfaces will accelerate comparative analysis of QTL and GWAS loci across the plant species, which will also contribute to gene isolation and molecular breeding.

ACKNOWLEDGEMENTS

We thank Ms. Mitsuyo Kohara for her technical assistance. This work was supported by the KAKENHI Grant-in-Aid for Scientific Research (C) (24510286), Japan Society for the Promotion of Science; and the Kazusa DNA Research Institute Foundation.

REFERENCES

1. Arumuganathan K. and Earle E.D. (1991) Nuclear DNA content of some important plant. Plant Mol. Biol. Rep. 9: 208–218
2. Asamizu E., Ichihara H., Nakaya A., Nakamura Y., Hirakawa H., Ishii T., Tamura T., Fukami-Kobayashi K., Nakajima Y. and Tabata S. (2014) Plant Genome DataBase Japan (PGDBj): a portal website for the integration of plant genome-related databases. Plant Cell Physiol. 55: e8.
3. CBOL Plant Working Group (2009) A DNA barcode for land plants. Proc. Natl. Acad. Sci. USA 106: 12794–12797
4. Fang Z., Polacco M., Chen S., Schroeder S., Hancock D., Sanchez H. and Coe E. (2003) cMap: the comparative genetic map viewer. Bioinformatics 19: 416–417
5. Gusella J.F., Wexler N.S., Conneally P.M., Naylor S.L., Anderson

M.A., Tanzi R.E., Watkins P.C., Ottina K., Wallace M.R., Sakaguchi A.Y.et al. (1983) A polymorphic DNA marker genetically linked to Huntington's disease. Nature 306: 234–238

6. Hamilton J.P. and Buell C.R. (2012) Advances in plant genome sequencing. Plant J. 70: 177–190
7. Handberg K. and Stougaard J. (1992) *Lotus japonicus*, an autogamous, diploid legume species for classical and molecular genetics. Plant J. 2: 487–496
8. Hayashi M., Miyahara A., Sato S., Kato T., Yoshikawa M., Taketa M., Hayashi M., Pedrosa A., Onda R., Imaizumi-Anraku H.et al. (2001) Construction of a genetic linkage map of the model legume *Lotus japonicus* using an intraspecific F_2 population. DNA Res. 8: 301–310
9. Helentjaris T., Slocum M., Wright S., Schaefer A. and Nienhuis J. (1986) Construction of genetic linkage maps in maize and tomato using restriction fragment length polymorphisms. Theor. Appl. Genet. 72: 761–769
10. Hirakawa H., Nakamura Y., Kaneko T., Isobe S., Sakai H., Kato T., Hibino T., Sasamoto S., Watanabe A., Yamada M.et al. (2011) Survey of the genetic information carried in the genome of *Eucalyptus camaldulensis*. Plant Biotechnol. 28: 471–480
11. Hirakawa H., Shirasawa K., Ohyama A., Fukuoka H., Aoki K., Rothan C., Sato S., Isobe S. and Tabata S. (2013a) Genome-wide SNP genotyping to infer the effects on gene functions in tomato. DNA Res.20: 221–233
12. Hirakawa H., Shirasawa K., Kosugi S., Tashiro K., Nakayama S., Yamada M., Kohara M., Watanabe A., Kishida Y., Fujishiro T.et al. (2013b) Dissection of the octoploid strawberry genome by deep-sequencing of the genomes of *Fragaria* species. DNA Res. 2: 169–181
13. Hisano H., Sato S., Isobe S., Sasamoto S., Wada T., Matsuno A., Fujishiro T., Yamada M., Nakayama S., Nakamura Y.et al. (2007) Characterization of the soybean genome using EST-derived microsatellite markers. DNA Res. 14: 271–281
14. International Rice Genome Sequencing Project (2005) The map-based sequence of the rice genome.Nature 436: 793–800
15. Isobe S., Klimenko I., Ivashuta S., Gau M. and Kozlov N.N.

(2003) First RFLP linkage map of red clover (*Trifolium pratense* L.) based on cDNA probes and its transferability to other red clover germplasm.Theor. Appl. Genet. 108: 105–112

16. Isobe S., Kölliker R., Hisano H., Sasamoto S., Wada T., Klimenko I., Okumura K. and Tabata S. (2009)Construction of a consensus linkage map for red clover (*Trifolium pratense* L.). BMC Plant Biol. 9: 57.
17. Isobe S.N., Hisano H., Sato S., Hirakawa H., Okumura K., Shirasawa K., Sasamoto S., Watanabe A., Wada T., Kishida Y.et al. (2012) Comparative genetic mapping and discovery of linkage disequilibrium across linkage groups in white clover (*Trifolium repens* L.). G3 2: 607–617
18. Isobe S.N., Hirakawa H., Sato S., Maeda F., Ishikawa M., Mori T., Yamamoto Y., Shirasawa K., Kimura M., Fukami M.et al. (2013) Construction of an integrated high density simple sequence repeat linkage map in cultivated strawberry (*Fragaria* × *ananassa*) and its applicability. DNA Res. 20: 79–92
19. Kofler R., Schlotterer C. and Lelley T. (2007) SciRoKo: a new tool for whole genome microsatellite search and investigation. Bioinformatics 23: 1683–1685
20. Koilkonda P., Sato S., Tabata S., Shirasawa K., Hirakawa H., Sakai H., Sasamoto S., Watanabe A., Wada T., Kishida Y.et al. (2012) Large-scale development of expressed sequence tag-derived simple sequence repeat markers and diversity analysis in *Arachis* spp. Mol. Breed. 30: 125–138
21. Kumar L.S. (1999) DNA markers in plant improvement: an overview. Biotechnol. Adv. 17: 143–182
22. Michael T.P. and Jackson S. (2013) The first 50 plant genomes. Plant Genome 6: 10.3835/plantgenome2013.03.0001in. Cross Ref
23. Moscone E.A., Baranyi M., Ebert I., Greilhuber J., Ehrendorfer F. and Hunziker A.T. (2003) Analysis of nuclear DNA content in *Capsicum* (Solanaceae) by flow cytometry and Feulgen densitometry. Ann. Bot. 92: 21–29
24. Phillips R.L. and Vasil I.K. (2001) DNA-based markers in plants, 2nd ed Kluwer Academic Publishers, Dordrecht, p. 512
25. Rozen S. and Skaletsky H. (2000) Primer3 on the WWW for

general users and for biologist programmers.Methods Mol. Biol. 132: 365–386

26. Sato S., Kaneko T., Nakamura Y., Asamizu E., Kato T. and Tabata S. (2001) Structural analysis of a*Lotus japonicus* genome. I. Sequence features and mapping of fifty-six TAC clones which cover the 5.4 Mb regions of the genome. DNA Res. 8: 311–318
27. Sato S., Isobe S., Asamizu E., Ohmido N., Kataoka R., Nakamura Y., Kaneko T., Sakurai N., Okumura K., Klimenko I.et al. (2005) Comprehensive structural analysis of the genome of red clover (*Trifolium pratense* L.). DNA Res. 12: 301–364
28. Sato S., Nakamura Y., Kaneko T., Asamizu E., Kato T., Nakao M., Sasamoto S., Watanabe A., Ono A., Kawashima K.et al. (2008) Genome structure of the legume, *Lotus japonicus*. DNA Res. 15: 227–239
29. Schmutz J., Cannon S.B., Schlueter J., Ma J., Mitros T., Nelson W., Hyten D.L., Song Q., Thelen J.J., Cheng J.et al. (2010) Genome sequence of the palaeopolyploid soybean. Nature 463: 178–183
30. Shirasawa K., Asamizu E., Fukuoka H., Ohyama A., Sato S., Nakamura Y., Tabata S., Sasamoto S., Wada T., Kishida Y.et al. (2010a) An interspecific linkage map of SSR and intronic polymorphism markers in tomato. Theor. Appl. Genet. 121: 731–739
31. Shirasawa K., Isobe S., Hirakawa H., Asamizu E., Fukuoka H., Just D., Rothan C., Sasamoto S., Fujishiro T., Kishida Y.et al. (2010b) SNP discovery and linkage map construction in cultivated tomato. DNA Res. 17: 381–391
32. Shirasawa K., Oyama M., Hirakawa H., Sato S., Tabata S., Fujioka T., Kimizuka-Takagi C., Sasamoto S., Watanabe A., Kato M.et al. (2011) An EST-SSR linkage map of *Raphanus sativus* and comparative genomics of the Brassicaceae. DNA Res. 18: 221–232
33. Shirasawa K., Hirakawa H., Tabata S., Hasegawa M., Kiyoshima H., Suzuki S., Sasamoto S., Watanabe A., Fujishiro T. and Isobe S. (2012a) Characterization of active miniature inverted-repeat transposable elements in the peanut genome. Theor. Appl. Genet. 124: 1429–1438
34. Shirasawa K., Koilkonda P., Aoki K., Hirakawa H., Tabata S., Watanabe M., Hasegawa M., Kiyoshima H., Suzuki S.,

Kuwata C.et al. (2012b) *In silico* polymorphism analysis for the development of simple sequence repeat and transposon markers and construction of linkage map in cultivated peanut. BMC Plant Biol. 12: 80.

35. Shirasawa K., Bertioli D.J., Varshney R.K., Moretzsohn M.C., Leal-Bertioli S.C., Thudi M., Pandey M.K., Rami J.F., Fonceka D., Gowda M.V.et al. (2013a) Integrated consensus map of cultivated peanut and wild relatives reveals structures of the A and B genomes of *Arachis* and divergence of the legume genomes. DNA Res. 20: 173–184
36. Shirasawa K., Fukuoka H., Matsunaga H., Kobayashi Y., Kobayashi I., Hirakawa H., Isobe S. and Tabata S. (2013b) Genome-wide association studies using single nucleotide polymorphism markers developed by re-sequencing of the genomes of cultivated tomato. DNA Res. 20: 593–603
37. Shirasawa K., Ishii K., Kim C., Ban T., Suzuki M., Ito T., Muranaka T., Kobayashi M., Nagata N., Isobe S.et al. (2013c) Development of *Capsicum* EST-SSR markers for species identification and *in silico*mapping onto the tomato genome sequence. Mol. Breed. 31: 101–110
38. Shirasawa K. and Hirakawa H. (2013) DNA marker applications to molecular genetics and genomics in tomato. Breed. Sci. 63: 21–30
39. Shulaev V., Sargent D.J., Crowhurst R.N., Mockler T.C., Folkerts O., Delcher A.L., Jaiswal P., Mockaitis K., Liston A., Mane S.P.et al. (2011) The genome of woodland strawberry (*Fragaria vesca*). Nat. Genet. 43: 109–116
40. Stein L.D., Mungall C., Shu S., Caudy M., Mangone M., Day A., Nicherson E., Stajich J.E., Harris T.W., Arva A.et al. (2002) The generic genome browser: a building block for a model organism system database. Genome Res. 12: 1599–1610
41. Temnykh S., DeClerck G., Lukashova A., Lipovich L., Cartinhour S. and McCouch S. (2001)Computational and experimental analysis of microsatellites in rice (*Oryza sativa* L.): frequency, length variation, transposon associations, and genetic marker potential. Genome Res. 11: 1441–1452

42. The Arabidopsis Genome Initiative (2000) Analysis of the genome sequence of the flowering plant*Arabidopsis thaliana*. Nature 408: 796–815
43. The *Brassica rapa* Genome Sequencing Project Consortium (2011) The genome of the mesopolyploid crop species *Brassica rapa*. Nat. Genet. 43: 1035–1039
44. The Tomato Genome Consortium (2012) The tomato genome sequence provides insights into fleshy fruit evolution. Nature 485: 635–641
45. ThielT., MichalekW., VarshneyR.K. and GranerA. (2003) Exploiting EST databases for the development and characterization of gene-derived SSR-markers in barley (*Hordeum vulgare* L.).Theor. Appl. Genet. 106: 411–422
46. Wu F. and Tanksley S.D. (2010) Chromosomal evolution in the plant family Solanaceae. BMC Genomics11: 182.

Chapter 9

Pre-Breeding for Diversification of Primary gene pool and Genetic Enhancement of Grain Legumes

Shivali Sharma, H. D. Upadhyaya, R. K. Varshney, and C. L. L. Gowda

International Crops Research Institute for the Semi-Arid Tropics, Hyderabad, India

ABSTRACT

The narrow genetic base of cultivars coupled with low utilization of genetic resources are the major factors limiting grain legume production and productivity globally. Exploitation of new and diverse sources of variation is needed for the genetic enhancement of grain legumes. Wild relatives with enhanced levels of resistance/tolerance to multiple stresses provide important sources of genetic diversity for crop improvement. However, their exploitation for cultivar improvement is limited by cross-incompatibility barriers and linkage drags. Pre-breeding provides a unique opportunity, through the introgression of desirable genes from wild germplasm into genetic backgrounds readily used by the breeders with minimum linkage drag, to overcome this. Pre-breeding activities using promising landraces, wild relatives, and popular cultivars have been initiated at International Crops Research Institute for the Semi-Arid Tropics (ICRISAT) to develop new gene pools in chickpea, pigeonpea, and groundnut with a high frequency of useful genes, wider adaptability, and a broad genetic base. The availability of molecular markers will greatly assist in reducing linkage drags and increasing the efficiency of introgression in pre-breeding programs.

INTRODUCTION

Grain legumes are second in importance to human and animal diets after cereals and occupy an important place in the world's food and nutrition economy. Being the primary source of high quality dietary protein, grain legumes are important in alleviating protein deficiency and malnutrition prevailing among poor people in developing countries, as well as contributing significantly to global food and nutritional security. Further, grain legumes provide high quality nutritious fodder for animal consumption. In addition to food and fodder, the nitrogen fixing capacity of grain legumes decreases the need for direct application of N-fertilizers and makes them an important component in cropping systems for improving and sustaining soil fertility and texture (Graham and Vance, 2003).

From the nutritional viewpoint, grain legumes are usually deficient in sulphur-containing amino acids-methionine, cysteine and tryptophan, but are rich in other essential amino acid, lysine. The reverse is true with cereals (Rockland and Radke, 1981). Therefore, the dietary mixture of cereals with legumes constitutes a source of balanced diet to the poor farmers in semi-arid tropical regions. In the past, the staple cereal crops, especially wheat, rice and maize, have received the highest research priority and considerable yield improvements were made in these crops. In contrast, grain legumes are under researched compared to cereals. Low yield potential coupled with biotic and abiotic stresses has further reduced their cultivation by the farmers. Recently, realizing the significance of grain legumes in improving nutrition and the livelihood of poor farmers, more research is now being carried out for their genetic amelioration by various institutes. The International Crops Research Institute for the Semi-Arid Tropics (ICRISAT), Patancheru, India is working for the genetic improvement of three major grain legumes, i.e., chickpea (*Cicer arietinum* L.), pigeonpea (*Cajanus cajan* Millsp.), and groundnut (*Arachis hypogaea* L.).

Grain legumes are cultivated mostly in marginal lands under rainfed conditions, with low and unstable productivity (Kumar and van Rheenen, 2000). Their production is adversely affected by several biotic and abiotic stresses. *Ascochyta* blight, *Botrytis* gray mold, *Fusarium* wilt, and dry root rot in chickpea;*Phytophthora* blight, and sterility mosaic disease in pigeonpea; and early and late leaf spots (LLSs), rust and aflatoxin contamination in groundnut are the most important and widely distributed diseases affecting yield and quality. Besides, pod borer (*Helicoverpa armigera* L.), drought, heat, salinity, water-logging are the other important stresses potentially limiting their productivity worldwide. Grain legumes, including chickpea, pigeonpea and groundnut have a narrow genetic base, due to the bottlenecks associated with their evolution and domestication, as well as due to the replacement of locally adapted crop landraces by the genetically advanced modern varieties. Low grain legume productivity due to biotic/abiotic stresses coupled with limited genetic variation in the cultivated gene pool necessitates the identification and utilization of diverse germplasm sources to develop new high-yielding cultivars with a broad genetic base.

GRAIN LEGUMES GENETIC RESOURCES

Plant genetic resources are reservoirs of natural genetic variation and provide raw material for crop improvement programs. About 7.4 million germplasm accessions of different crops have been collected and/or assembled and conserved in over 1750 *ex situ* genebanks worldwide (FAO, 2010) and the grain legumes constitute the second largest collection (15%) after cereals (45%). Globally, ~1.1 million grain legume germplasm accessions are conserved in different genebanks, of which ICRISAT genebank holds ~50,000 accessions including cultivated and wild relatives of chickpea (20,267 accessions), pigeonpea (13,771 accessions), and groundnut (15,445 accessions) from 133 countries. Besides ICRISAT, other major genebanks holding grain legume germplasm are the National Bureau of Plant Genetic Resources (NBPGR), New Delhi, India (16,881 *Cicer*, 12,900 *Cajanus* and 14,593*Arachis* accessions); the International Centre for Agricultural Research in Dry Areas (ICARDA), Aleppo, Syria (13,818 *Cicer* accessions); the United States Department of Agriculture (USDA), USA (9964 *Arachis* accessions); and the Directorate of Groundnut Research (DGR), Junagadh, India (8934*Arachis* accessions) (updated report based on species inhttp://apps3.fao.org/wiews/germplasm_query.htm?i_l=EN).

At present, emphasis is on conserving accessions safely with duplication both within and outside the CGIAR. At global level, the Svalbard Global Seed Vault has been established in the Norwegian island of Spitsbergen to store seed samples of a wide variety of plant species in an underground cavern. This seed vault will provide an insurance against the loss of seeds in genebanks, as well as a refuge for seeds in the case of large-scale regional or global crises. Over the last 5 years, >780,000 seed samples belonging to about 840 genera have been deposited in this vault of which about 170,000 accessions belong to 19 legume genera/species (http://www.nordgen.org/sgsv/index.php?app=data_unit&unit= sgsv_by_species&PHPSESSID=1hoc3n9kkdmng79o5vlcmsgs25).

USE OF GERMPLASM IN GRAIN LEGUME IMPROVEMENT

Surprisingly, in spite of large collections, only a few germplasm accessions (<1%) have been utilized in crop improvement programs such as in wheat (Dalrymple, 1986), maize (Cantrell et al., 1996), spring barley (Vellve, 1992), soybean (Mikel et al., 2010), and other grain legumes (Kumar et al.,2004). India has one of the largest grain legume breeding programs and has released about 230 cultivars of chickpea, pigeonpea, lentil, black gram, and green gram through hybridization and selection (data up to 2003). Pedigree analysis of these cultivars revealed that Pb-7 in chickpea, T-1 and T-190 in pigeonpea, L-9-12 in lentil, T-9 in black gram, and T-1 in green gram were the most frequently used parents (Kumar et al., 2004). In ICRISAT, which has the largest germplasm collections of chickpea, pigeonpea and groundnut, the breeding programs have used only 91 chickpea (0.4% of total 20,267 accessions), 54 pigeonpea (0.4% of total 13,771 accessions) and 171 groundnut (1.1% of total 15,445 accessions) germplasm accessions to develop advanced breeding lines. Further, the increase in accession numbers in genebanks and lack of corresponding increase in their utilization by the crop improvement scientists also indicate that the collections are not being utilized to their full potential (Marshall, 1989). As a global responsibility, ICRISAT's genebank has supplied over 303,000 samples of grain legumes, which includes 131,924 samples of chickpea, 71,826 samples of pigeonpea, and 99,325 samples of groundnut accessions to scientists across 136 countries. Besides this, over 370,000 samples of these legumes have also been distributed to researchers within ICRISAT during 1974–June 2013. Some of the germplasm accessions supplied from ICRISAT genebank have been released directly as varieties by NARS such as 15 chickpea germplasm accessions released as varieties in 15 countries, 10 pigeonpea germplasm accessions in seven countries and 11 groundnut germplasm accessions in 15 countries (Table 1). These varieties have greatly benefited the farmers by contributing to increase in production and productivity in these countries. However, the pattern of demand and consequent supply has shown a greater demand for a few specific germplasm accessions. In a period of 35 years, in chickpea, two accessions, ICC 4918 and ICC 4973 were supplied over 350 times, and four accessions more than 200 times. In pigeonpea, one accession, ICP 7035 was supplied 330 times, and seven accessions more than 200 times. Similarly, in groundnut, one accession, ICG 799 was supplied over 300 times, and five accessions about 200 times. Further, 3740 chickpea, 3088 pigeonpea and 1018 groundnut accessions have not been requested at all, while 7339 chickpea, 7001 pigeonpea and 7582 groundnut accessions have been supplied less than five times (as on 30 May 2013).

Table 1. List of chickpea, pigeonpea and groundnut germplasm accessions directly released as varieties for cultivation

Accession number	Country of origin	Year of release	Country of release	Released name
CHICKPEA				
ICC 237	India	1988	Oman	ICC 237
ICC 552	India	–	Myanmar	Yezin 1
ICC 3274	Iran	1999	Bangladesh	Bari Chhola7
ICC 4923	India	1978	India	Jyothi
ICC 4944	India	–	Myanmar	Keyhman
ICC 4951	India	–	Myanmar	ICC 4951
ICC 4998	India	1994	Bangladesh	Bina-Sola 2
ICC 6098	India	1987	Nepal	Radha
ICC 8521	Italy	–	USA	Aztee
ICC 8649	Afghanistan	1987	Sudan	Shendi
ICC 11879	Turkey	1986	Turkey	–
	Turkey	1982	Syria	Ghab 1
	Turkey	1987	Morocco	–
	Turkey	1988	Algeria	–
ICC 13816	USSR (former)	1986	Syria	Ghab 2
	USSR (former)	1984	Algeria	Yialousa
	USSR (former)	–	Cyprus	–
	USSR (former)	1987	Italy	Sultano
ICC 14808	India	2006	Ethiopia	Yelbey
ICC 14880	India	1997	Australia	Hira
ICC 14911	USSR (former)	1987	Morocco	–
	USSR (former)	1986	Turkey	–
PIGEONPEA				
ICP 6997	India	1992	Nepal	Rampur Rhar
ICP 7035	India	1989	Philippines	–
	India	2005	India	JK Sweety (JKPL 5)
	India	2003	China	Guimu 4
	India	1985	Fiji	Kamica
ICP 8863	India	1985	India	Maruti

ICP 9145	Kenya	1988	Malawi	Nandolo wa nswana
ICP 9905	India	1991	Venezuela	La Cerrera
ICP 11384	Nepal	1992	Nepal	Bageswari
ICP 11916	India	1991	Venezuela	Aroa
ICP 13092	Kenya	2005	India	JK Sixer (JKPL 6)
ICP 13829	Granada	1991	Venezuela	Cerro Pelon
ICP 14770	India	1989	India	Abhaya
GROUNDNUT				
ICG 221	India	1994	Swaziland	–
ICG 273	Argentina	1994	Ethiopia	Sedi
ICG 1697	Peru	1998	Indonesia	Singa
ICG 1703	Peru	1998	Indonesia	Panter
	Peru	2003	Thailand	Kalasin 2
ICG 2271	USA	–	Nepal	–
ICG 2974	Israel	1984	Myanmar	Sinpadetha 3
	Israel	1985	Tanzania	Johri
ICG 7794	USA	1989	Ethiopia	–
ICG 7827	India	1992	Philippines	UPL Pn 10
	India	1984	Myanmar	Sinpadetha 2
ICG 7878	India	–	Mali	Waliyartiga
	India	1990	Mauritius	ICG 7878
	India	2011	Mozambique	JL24
ICG 7886	Peru	1987	Jamica	Cardi-Payne
ICG 12991	India	2004	Zambia	Msandile
	India	2001	Malawi	Baka
	India	2002	Uganda	Serenut 4T
	India	2002	Mozambique	Nametil

REASONS FOR LOW USE OF GERMPLASM

The factors responsible for low use of germplasm in crop improvement programs are the large size of germplasm collections, the breeders' preference for working collections, and the linkage drag associated with utilizing wild relatives in crop improvement programs. Large germplasm collections of most crops, including grain legumes, poses problem for their meaningful multi-locational evaluation. Thus, there is a lack of information on traits of economic importance, which often shows high genotype × environment interaction, for large number of germplasm accessions. This creates a problem for the breeders to select the appropriate genetic diversity for use in their breeding programs. However, strategies are now available to overcome the problems associated with the large size of germplasm collections. This includes the development of representative small-sized subsets such as core (Frankel and Brown, 1984) and mini core (Upadhyaya and Ortiz, 2001) collections that contain the majority of allelic diversity. Besides large size, there are apprehensions among breeders about poor adaptability of germplasm and linkage drag. Linkage drag is the most important factor responsible for low use of germplasm in crop improvement and is the major reason for the need for pre-breeding. While using unknown and wild germplasm, comparatively more efforts, time and resources are required to break undesirable linkage drag during the development process, particularly for regional adaptability to climates, crop management, biotic and abiotic stresses, and overall agronomic performance. This makes the breeding program comparatively more lengthy and cumbersome (Figure (Figure1).1). For these reasons, breeders use their working collection in breeding programs which results in re-circulation of same genotype and hence narrow genetic base of the released cultivars. Nevertheless, the above reservations and over cautiousness toward linkage drag has resulted in continuous narrowing down of genetic variability among modern cultivars, with implications for survival from several biotic and abiotic stresses.

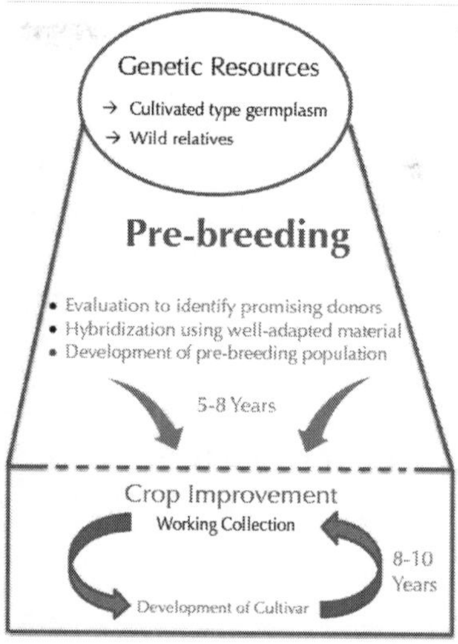

Figure 1. Pre-breeding as a bridge between genetic resources and crop improvement.

PRE-BREEDING FOR ACCESSING NOVEL GENES

The success of any crop improvement program depends on the availability of sufficient genetic variability, but this variability must be in conventionally usable form. The variability available in any crop germplasm conserved in genebanks for present and future use belongs broadly to the following three groups:

- Cultivated type
- Cross-compatible wild type
- Cross-incompatible wild type

The genetic variability in cultivated type germplasm is either in poor agronomic background or in genetic background not adapted to the breeding or target climate for its direct use in conventional breeding programmes. The exploitation of genetic variability in wild species for cultivar improvement is hindered mainly by linkage drag and different incompatibility barriers between cultivated and wild species. Under such situations, pre-breeding offers a unique tool to enhance the use of genetic variability present both in cultivated and wild type germplasm.

Pre-breeding involves all the activities associated with identification of desirable traits and/or genes from unadapted germplasm (donor) that cannot be used directly in breeding populations (exotic/wild species), and to transfer these traits into well-adapted genetic backgrounds (recipients) resulting in the development of an intermediate set of material which can be used readily by the plant breeders in specific breeding programmes to develop new varieties with a broad genetic base ((Figure1).

Identification of trait-Specific Germplasm Accessions

Cultivated type Germplasm

Enormous efforts are needed to evaluate germplasm for traits of economic importance, as well as to screen for biotic and abiotic stress related traits through reliable and standardized techniques for identifying potential donors. Owing to the large size of germplasm collections, this becomes a costly and resource-demanding task. Following the concept of Frankel and Brown (1984) for sampling diversity, a number of small-sized subsets such as core (10% of the entire collection in size) and mini core (10% of core or 1% of entire collection) collections representing the diversity of entire collections have been developed in chickpea, pigeonpea and groundnut (Table 2). Due to their small size, these collections, especially mini core collections, provide an easy means to evaluate these collections across multilocations and to identify promising germplasm accessions as new and potential donors for various traits. Over 90 sets of mini core collections of SAT legumes have been shared by ICRISAT genebank with researchers in 20 countries. Evaluation of germplasm including core/mini core collections globally has resulted in the identification of new and diverse trait-specific germplasm accessions for agronomic and nutritional traits, as well as for resistance/tolerance to various biotic/abiotic stresses (reviewed in Upadhyaya et al., 2010). For example, new and promising sources for early maturity, large seed size, yield and component traits, and for nutrition-related traits, are now available for improvement of chickpea, pigeonpea and groundnut (Table 3). Similarly, for abiotic/biotic stresses, new sources of tolerance/resistance have been identified in chickpea, pigeonpea and groundnut (Table 4).4). For example, in chickpea, in addition to the well-known drought tolerant accession, ICC 4958, new sources of drought tolerance have been identified using different techniques (Table 4); ICC 13124 has the highest drought tolerance efficiency (DTE), lowest drought susceptibility index (DSI) and the highest harvest index (HI) and has been identified as the most drought tolerant accession (Parameshwarappa and Salimath, 2008; Parameshwarappa et al.,2010). Recently, to combat the effect of climate change on grain legume production, new and diverse sources of heat tolerance (Krishnamurthy et al., 2011a; Upadhyaya et al., 2011a), herbicide tolerance (Tar'an et al., 2010) and accessions with greater biological nitrogen fixation (BNF) capacity (Biabani et al., 2011) for chickpea and water-logging tolerant accessions for pigeonpea improvement (Krishnamurthy et al., 2011b) (Table 4) have also been identified.

Table 2. **Core and mini core collections as reported in chickpea, pigeonpea and groundnut**

Reduced subset	**Accessions used (No.)**	**Collection**	**Accessions in constituted**	**References**
			collection (No.)	
Chickpea	16,991	Core	1956	Upadhyaya et al., 2001a
	3350	Core	505	Hannan et al., 1994
	1002	Core	158	Kibret, 2012
	–	*Kabuli* core	103	Pouresmael et al., 2009
	1956	Mini core	211	Upadhyaya and Ortiz, 2001
	482	Mini core	39	Biabani et al., 2011
Pigeonpea	12,153	Core	1290	Reddy et al., 2005
	1290	Mini core	146	Upadhyaya et al., 2006a

Groundnut	630	Valencia core	77	Dwivedi et al., 2008b
	7432	Core	831	Holbrook et al., 1993
	6390	Core	576	Jiang et al., 2008
	4738	Asian core	504	Upadhyaya et al., 2001b
	14,310	Core	1704	Upadhyaya et al., 2003
	831	Mini core	111	Holbrook and Dong, 2005
	1704	Mini core	184	Upadhyaya et al., 2002

Table 3. **Promising germplasm accessions for agronomic and nutrition-related traits in chickpea, pigeonpea and groundnut**

Traits	**Chickpea**	**Pigeonpea**	**Groundnut**
Early maturity	28 accession (ICC# 16641, 16644, 11040, 11180, and 12424; Upadhyaya et al., 2007a)*	8 accession (ICP# 14900, 1156, 14471, 14903, and 16309; Upadhyaya et al., 2010)	21 accession (ICG# 4558, 4890, 9930, 11605, and 5512; Upadhyaya et al., 2006b)
Large seed	11 accessions (ICC# 14190, 14194, 7345, 17452, 19189, and 17109; Gowda et al., 2011)	3 accessions (ICP# 14976, 13359, and 13139; Upadhyaya et al., 2010)	12 accessions (ICG# 2381, 5016, 5051, 5745, and 5662; Upadhyaya et al., 2010)
	4 accessions (ICC# 14199, 14197, 14203, and 12033; Kaul et al., 2007)		
	9 accessions (ICC# 17457, 17452, 19189, 17456, and 18591; Kashiwagi et al., 2007)		

Yield and component traits	39 accessions (ICC# 6122, 8474, 8155, 12034, and 4871; Upadhyaya et al., 2007b)	4 accessions (ICP# 8860, 11230, 4167, and 8602; Upadhyaya et al., 2010)	60 accessions (ICG# 4, 29, 3443, 14161, 11188, 7140, and 2918; Upadhyaya et al., 2005)
	8 accessions (ICC# 13124, 12654, 9848, 6279, 5879, and 10341; Parameshwarappa et al., 2011a,b)		
	6 accessions (ICC# 14778, 6279, 4567, 4533, 1397, and 12328; Meena et al., 2010)		
Protein content (>30%)	–	6 accessions (ICP# 4575, 7426, 8266, 11823, 12515 and 12680; Upadhyaya et al., 2013)	18 accessions (ICG # 36, 5779, 3421, 3584, and 2019; Upadhyaya et al., 2012b)
Oil content (>50%)	–	–	18 accessions (ICG # 442, 397, 5779, 4955 and 14710; Upadhyaya et al., 2012b)
Oleic acid (≥60%)	–	–	18 accessions (ICG # 6022, 12625 and 11088 in subsp.*fastigiata*; and ICG# 2381, 10185, 15419, 12276, 7243 and 6766 in subsp. *hypogaea*; Upadhyaya et al., 2012b)
Oleic/ Linoleic acid ratio (>3.0)	–	–	18 accessions (ICG # 6022, 12625 and 1274 in subsp.*fastigiata*; and ICG# 2381, 10185, 15419, 12276, 7243 and 6766 in subsp. *hypogaea*; Upadhyaya et al., 2012b)

Table 4. **New sources of tolerance/resistance to abiotic/biotic stresses for chickpea, pigeonpea and groundnut improvement**

Traits	Promising germplasm	References
CHICKPEA		
Drought	Root length density: 9 accessions (ICC# 8261, 10885, 16796, 13816 and 13599)*	Kashiwagi et al.,2005
	Root depth: 9 accessions (ICC# 3512, 15697, 13523, 1356, and 4872)	Kashiwagi et al.,2005
	SPAD Chlorophyll Meter Reading (SCMR): 4 accessions (ICC# 16374, 1422, 16903, and 10945)	Kashiwagi et al.,2006, 2010
	Canopy temperature: 3 accessions (ICC# 3325, 867, and 14799)	Kashiwagi et al.,2008
	Terminal drought: 5 accessions (ICC# 867, 1923, 9586, 12947 and 14778)	Krishnamurthy et al.,2010
Salinity	10 accessions (ICC# 10755, 13124, 13357, 15406, and 15697)	Serraj et al., 2004
	15 accessions (5003, 15610, 1431, 4593, and 12155)	Vadez et al., 2007
	12 accessions (9942, 6279, 11121, 456 and 14799)	Krishnamurthy et al.,2011c
Heat	18 accessions (ICC# 456, 637, 1205, 3362, and 3761)	Krishnamurthy et al.,2011a
	10 accessions (14346, 5829, 6121, 13124 and 14284)	Upadhyaya et al.,2011a
Herbicide (imazethapyr and imazamox)	4 accessions (ICC# 2242, 2580, 3325 and ILC 4279)	Tar'an et al., 2010
Multiple disease	18 accessions (ICC# 1710, 2242, 2990, 11284, and 12328)	Pande et al., 2006
PIGEONPEA		
Water-logging	24 accessions (ICP# 1279, 7057, 7148, 10397, and 16309)	Krishnamurthy et al.,2011b
Salinity	16 accessions (ICP# 8860, 7426, 14722, 11477, and 6128)	Srivastava et al.,2006

Combined resistance for *Fusarium* wilt and sterility mosaic	6 accessions (ICP# 6739, 8860, 11015, 13304, 14638, and 14819)	Sharma et al., 2012
GROUNDNUT		
Drought	18 accessions (ICG# 14523, 6766, 7243, 862, and 6654)	Upadhyaya, 2005
	30 accessions (11088, 12697, 8751, 3140, and 3584)	Hamidou et al., 2011
Low temperature	15 accessions (ICG# 12625, 7898, 11130, 6148, and 7013)	Upadhyaya et al.,2009
Salinity	14 accessions (ICG# 5195, 442, 7283, 1711, and 2106)	Srivastava, 2010
Rhizoctonia limb rot resistant	6 accessions (PI# 343398, 343361, 288178, 331326, 497351, and 274193)	Franke et al., 1999
Late leaf spot	7 accessions (ICG# 12625, 11426, 12672, 13787, 14475, 2857, and 8760)	Kusuma et al., 2007
Rust	22 accessions (ICG 9809, 11088, 11426, 13787, and 8760)	Kusuma et al., 2007
A. flavus	6 accessions (ICG# 14985, 3673, 6025, 12625, 13787, and 8760),	Kusuma et al., 2007
	20 accessions (6813, 12370, 4750, 4156, and 12697)	Jiang et al., 2010
Bud necrosis	4 accessions (ICG# 875, 928, 1668, and 14466)	Ahmed, 2008
Combined resistance to *Sclerotinia* blight, pepper spot and web blotch	5 accessions (PI# 274193, 497599, 458619, 468195, and 259796)	Damicone et al.,2010

Wild Type Germplasm

Wild species are the reservoir of many useful genes/alleles as they have evolved under natural selection to survive climate extremes. Wild species of *Cicer, Cajanus*, and *Arachis* have been extensively screened and several of them were reported to have very high level of resistance/tolerance to various stresses. Among wild *Cicer* species, *C. bijugum, C. judaicum*, and *C. pinnatifidum* are the most important sources having the highest levels of resistance/tolerance to multiple stresses (reviewed in Gaur et al., 2009; Sharma et al., 2013). Wild *Cajanus* species especially,*C. scarabaeoides, C. acutifolius, C. platycarpus, C. reticulatus, C. sericeus*, and *C. albicans* provide resistance to pod borer, *Helicoverpa armigera* (Rao et al., 2003; Sujana et al., 2008; Sharma et al.,2009). Evaluation of wild *Cajanus* species has identified new sources of resistance to *Alternaria* blight,*Phytophthora* blight, sterility mosaic virus, pod fly and nematodes and tolerance to salinity, drought, and photoperiod insensitivity (reviewed in Upadhyaya et al., 2013). Good sources of resistance for bruchids (*Callosobrochus maculatus*) have also been identified in *C. scarabaeoides, C. acutifolius*, and*C. platycarpus* (Jadhav et al., 2012). Similarly, wild *Arachis* species harbor very high levels of resistance/tolerance to many biotic/abiotic stresses (reviewed in Dwivedi et al., 2008a; Upadhyaya et al., 2012a). These wild species are the potential donors to develop genome-wide introgression (GWI) lines for the genetic amelioration of chickpea, pigeonpea and groundnut.

Pre-breeding for Grain Legume Genetic Enhancement

The success of any pre-breeding program depends mainly upon three factors: (1) identification of promising donor with good expression of the trait; (2) type of germplasm (Cultivated/cross-compatible wild type/cross-incompatible wild type); and (3) agronomic performance of the donors. New and diverse sources of variation for agronomic and nutrition-related traits and resistant/tolerant sources for biotic/abiotic stresses are now available both in cultivated and wild type germplasm and can be utilized to develop new pre-breeding populations having greater variability for various traits. In the past, a few promising wild type accessions have been utilized by some researchers for the improvement of chickpea, pigeonpea and groundnut and are given hereunder.

Pre-Breeding for Biotic/Abiotic Stresses

Chickpea

Chickpea Of the eight annual wild *Cicer* species, only *C. reticulatum* is readily crossable with cultivated chickpea resulting in fertile hybrid. The exploitation of the remaining seven annual wild *Cicer* species requires specialized techniques such as the application of growth hormones, embryo rescue, ovule culture, and tissue culture techniques (Badami et al., 1997; Mallikarjuna, 1999; Lulsdorf et al., 2005; Mallikarjuna and Jadhav, 2008). Utilization of *C. reticulatum* accession, ILWC 119 in crossing programme has resulted in the development of two cyst nematode resistant chickpea germplasm lines ILC 10765 and ILC 10766 (Malhotra et al., 2002). Beneficial traits such as cold tolerance and a high degree of resistance to wilt, foot rot, root rot, and*Botrytis* gray mold have also been introgressed from *C. reticulatum* and *C. echinospermum* into cultivated chickpea (ICARDA, 1995; Singh et al., 2005; Ramgopal et al., 2012). Using novel techniques, interspecific hybrids have been produced between *C. arietinum* × *C. judaicum* (Verma et al., 1990; Verma and RaviSandhu, 1995; Singh et al., 1999), *C. arietinum* × *C. pinnatifidum* (Verma et al., 1990; Badami et al., 1997; Mallikarjuna, 1999; Mallikarjuna and Jadhav, 2008), *C. arietinum* × *C. cuneatum* (Singh and Singh, 1989), and *C. arietinum* × *C. bijugum* (Verma et al., 1990; Singh et al.,1999; Mallikarjuna et al., 2007) to introgress desirable alien genes from these cross-incompatible wild*Cicer* species into cultivated chickpea. These interspecific hybrids have contributed significantly toward the development of genomic resources for chickpea improvement.

Pigeonpea

Wild *Cajanus* species, especially the cross-compatible secondary gene pool species, have been used for the genetic improvement of pigeonpea and the most significant achievement includes the development of unique cytoplasmic-nuclear male sterility system (CMS). The CMS systems have been developed with cytoplasm derived from wild *Cajanus* species namely *C. sericeus* (Ariyanayagam et al., 1995), *C. scarabaeoides* (Tikka et al., 1997; Saxena and Kumar, 2003), *C.*

volubilis (Wanjari et al.,2001), *C. cajanifolius* (Saxena et al., 2005), *C. lineatus* (Saxena et al., 2010), and *C. platycarpus*(Mallikarjuna et al., 2011a). For pigeonpea improvement, the potential for genetic introgression of salinity tolerance from wild relative, *Atylosia albicans* to cultivated pigeonpea has been demonstrated in the F_1 (Subbarao et al., 1990). Further, utilization of *C. acutifolius* as the pollen parent has resulted in the development of advanced generation population having resistance to pod borer (Mallikarjuna et al., 2007), variation in seed color and high seed weight. For *Phytophthora* blight, resistant genes have been transferred from *C. platycarpus* into cultivated pigeonpea and the resistant plants identified at seedling stage showed resistance across all stages of their life cycle (Mallikarjuna et al., 2005).

Groundnut

Utilization of wild *Arachis* species following interspecific hybridization has resulted in the development of many elite germplasm lines and cultivars with improved level of resistance to diseases and insect-pests. At ICRISAT, several elite lines have been developed with desirable characters transferred from wild *Arachis* species such as ICGV 86699 (Reddy et al., 1996) with multiple pest resistance, ICGV 87165 (Moss et al., 1998) with multiple disease and insect resistance, ICGV 99001 and 99004 with resistance to LLS, and ICGV 99003, and 99005 resistant to rust (Singh et al., 2003). Varieties such as Spancross (Hammons, 1970), Tamnut 74 (Simpson and Smith, 1975), Coan (Simpson and Starr, 2001), NemaTAM (Simpson et al., 2003), ICGV-SM 85048 (Nigam et al., 1998), and ICGV-SM 86715 (Moss et al., 1998), having a genetic base from wild *Arachis* species, were released for cultivation, mostly in USA. The development and utilization of synthetic amphiploids such as TxAG-6 with large genetic variation (Simpson et al., 1993) has made possible the transfer of resistance genes from wild species into cultivated groundnut. TxAG-6 is a synthetic amphiploid derived from crossing an AA-genome donor hybrid (*A. cardenasii* × *A. diogoi*) with a BB genome species, *A. batizocoi*, followed by colchicine treatment of the sterile triploid to produce fertile hexaploid, TxAG-6 (Simpson et al., 1993). This amphiploid has been synthesized using species that are not in the direct lineage of the cultivated groundnut. However, it is crossable with the cultivated groundnut and produced fertile progenies, thus proving useful for introducing genetic variability into the cultigen. Using this amphiploid in crossing programs with cultivated groundnut has resulted in the release of two cultivars, Coan and NemaTAM, carrying genes for root-knot nematode (*M. arenaria*) resistance from *A. cardenasii* (Simpson and Starr, 2001; Simpson et al., 2003).

Pre-Breeding for Improvement of Agronomic and Nutrition-Related Traits

Besides resistant sources, studies also indicated the possibility of improving agronomic traits, including yield, through introgression of genes from the wild species into the cultigens. In chickpea, by hybridizing *C. reticulatum* and *C. echinospermum* with cultigen, Singh and Ocampo (1997) generated high-yielding recombinant lines without any known undesirable traits from the wild species. In a similar study, introgressing wild genes from *C. reticulatum* into cultivated chickpea generated nine genotypes with high yield potential, resistance to soil-borne diseases and adaptation to water-limited environments. Three lines BG 1100, BG 1101, and BG 1103 produced 20% higher seed yield than the best-adapted cultivars and were resistant to *Fusarium* wilt (*Fusarium oxysporum* f.sp. *ciceris*) (Yadav et al., 2004). Interspecific derivatives possessing a high degree of resistance to diseases such as wilt, root rot and foot rot, and high yield, have been obtained from *C. arietinum* × *C. reticulatum* crosses (Singh et al.,2005). In ICRISAT, using two *C. reticulatum* accessions (110–113 days to 50% flowering, 143–150 days to maturity, 12–16 g 100-seed weight) in a hybridization program, several progenies were selected which took 8–21 days less to flower, 6–33 days less to mature, with 20–103% larger seeds, and 97–217% greater seed yield than respective cultivated (*C. arietinum*) parent (Upadhyaya, 2008). From *C. arietinum* × *C. judaicum* cross, a pre-breeding line IPC 71 having high number of primary branches, more pods per plant and green seeds has been developed for use in chickpea improvement programs (Chaturvedi and Nadarajan, 2010).

Further, utilization of wild *Cajanus* species has also contributed significantly toward the improvement of agronomic performance and nutritional quality of cultivated pigeonpea. A high protein line, ICPL 87162 was developed from the cross *C. cajan* × *C. scarabaeoides* (Reddy et al., 1997). This line contains 30–34% protein content compared to control cultivar (23% protein). Breeding lines with high protein content have also been developed from *C. sericeus*, *C. albicans*, and *C. scarabaeoides*. Utilization of wild *Cajanus* species has resulted in the development of several lines such as HPL 2, HPL 7, HPL 40, and HPL 51 having high protein and seed weight (Saxena et al., 1987). Recently, scientists at ICRISAT have generated segregants with up to 20.4 g 100-seed weight from the cross between cultivated pigeonpea, ICPL 85010 (10.5 g 100-seed weight) and *C. acutifolius*, ICP 15605 (2.2 g 100-seed weight) (Figure 2).

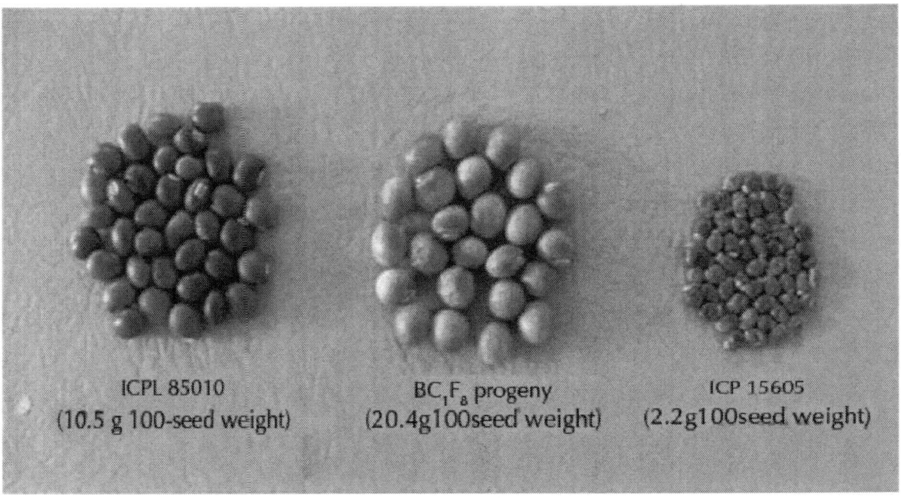

Figure 2. Improvement of seed traits in cultivated pigeonpea by using*Cajanus acutifolius*.

Similarly in groundnut, by using an amphiploid TxAG-6 [(*Arachis cardenasii* × *A. diogoi*) × *A. batizocoi*] with very low 100-seed weight (~12 g) and poor pod yield (2–5 g plant^{-1}), breeding lines have been developed. These have higher 100-seed weight (up to 95 g) compared to 32 g of control cultivar TMV 2 and 23–68% more pod yields than TMV 2 (3343 kg ha^{-1}). They also have 10–50% more pod yield than the high yielding control cultivar ICGV 91114 (3741 kg ha^{-1}, 49 g 100-seed weight^{-1}) (Upadhyaya, 2008). This demonstrates that the novel alleles of wild relatives that were considered to be lost in evolution to cultivated types could still be used to enhance the important agronomic and nutrition-related traits in cultivars.

PRE-BREEDING FOR CLIMATE SMART GRAIN LEGUME CROPS

The emerging threat to global crop production is climate variability, leading to frequent droughts as a result of erratic rainfall, high temperature prevalence, water-logging, increased soil salinity, and emergence of new insect-pests and diseases. Due to climate change, several areas are now becoming unfit for cultivation of traditional crops. To cope up with this situation, there is a need to breed new crop cultivars with a broad genetic base capable of withstanding frequent climatic fluctuations. In grain legumes, emerging threats due to climate change involves dry root rot (*Rhizoctonia bataticola*) and*Ascochyta* blight in chickpea and *Phytopthora* and *Alternaria* blight in pigeonpea. Frequent droughts coupled with high temperature during flowering to pod filling stage would result in heavy incidence of dry root rot whereas cooler temperature with high humidity are associated with high incidence of*Ascochyta* blight in chickpea (Pande et al., 2010). In pigeonpea, erratic and heavy rainfall (>300 mm in 6–7 days) leading to temporary flooding has resulted in an outbreak of *Phytophthora* blight in Deccan Plateau of India (Sharma et al., 2006). Recently, *Alternaria* blight of pigeonpea is becoming a serious threat in semi-arid tropics due to the untimely rainfall. Changing climatic conditions resulting in high rainfall will favor foliar diseases and some soilborne pathogens such as *Phytophthora, Pythium, Rhizoctonia solani,* and *Sclerotium rolfsii* whereas low rainfall resulting in decreased moisture will lead to the high incidence of *Fusarium* wilt, dry root rot etc. in cool season grain legumes (Pande and Sharma, 2011). Under such situations, pre-breeding activities utilizing wild species demand the highest priority and must focus on the identification of new traits better suited to the new environment for developing climate-resilient cultivars with a broad genetic base.

GENOMIC RESOURCES FOR USE IN PRE-BREEDING FOR GRAIN LEGUME IMPROVEMENT

Significant progress has been made in the development of large-scale genomic resources in the last decade for all the three SAT legume crops. Further, coordinated efforts have led to the development of large-scale genomic resources such as molecular markers, construction of comprehensive genetic maps, dense consensus maps and identification of marker-trait associations. Some recent reviews provide in-depth detail on quantity, source and development methods of these resources in these three legumes (see reviews Upadhyaya et al., 2011b; Varshney et al., 2012a, 2013a). Simple sequence repeat (SSR) or microsatellite and single nucleotide polymorphism (SNP) markers are the preferred marker systems especially for genetics and breeding applications. In addition, other genotyping systems such as diversity array technology (DArT) have also become available. A total of ~13,000–309,052 SSRs markers have become available in chickpea, pigeonpea and groundnut. ICRISAT, in collaboration with DArT Pty Ltd Australia, has developed DArT arrays with 15,360 features for chickpea (Thudi et al.,2011), groundnut and pigeonpea crops (see Varshney et al., 2013a). Although these DArT arrays showed low polymorphism among genotypes of the primary gene pool, they are of great help in monitoring the alien genome introgression in the cultivated species, as been observed for pigeonpea (Mallikarjuna et al., 2011b).

SNP markers are now gaining more importance due to higher abundance and amenability to high-throughput genotyping. Thousands of SNPs were identified using a variety of approaches in chickpea by Hiremath et al. (2011) (26,082 potential SNPs); in pigeonpea by Dubey et al. (2011) (12,141 SNPs) and Varshney et al. (2012a) (28,104 novel SNPs); and in groundnut by University of Georgia (8478 SNPs) (Nagy et al., 2012).

In addition to development of genomic tools essential for breeding applications, there is a great need to develop cost-effective genotyping platforms/assays. A range of cost-effective SNP genotyping platforms have become available such as Illumina GoldenGate assays for genotyping 768 SNPs in chickpea by University of California-Davis,

USA and 1536-SNPs for groundnut by University of Georgia, USA. In order to provide better genotyping deal for small to medium number of samples (<500), VeraCode assays have been developed for 96-plex and 48-plex SNP sets for chickpea and pigeonpea, respectively. Similarly, an alternative genotyping assay (KASPar assay) has been developed by KBiosciences (www.kbioscience.co.uk), which provides flexibility to genotype any number of samples with any number of SNPs. Thus, ICRISAT has developed KASPar assays for 2005 SNPs in chickpea (Hiremath et al., 2012), 1616 SNPs in pigeonpea (Saxena et al., 2012) and 90 SNPs in groundnut (Khera et al., 2013). Besides above mentioned genomic resources and genotyping platforms, draft genome sequences have become available in pigeonpea (Varshney et al., 2012b) and chickpea (Varshney et al., 2013b), while similar efforts are underway in groundnut. The recent availability of the above resources for these three crops has now changed their status from "genomics poor" to "genomics rich" crops.

PRE-BREEDING: PRESENT STATUS AND FUTURE OUTLOOK

Due to the limited genetic variability available in cultivated germplasm, pre-breeding is gaining importance in most crop improvement programs including wheat (Valkoun, 2001), maize (Nass and Paterniani, 2000), and common bean (Acosta-Gallegos et al., 2007). Recently, under the ICAR-ICARDA collaborative research, efforts are in progress for the improvement of bread wheat, kabuli chickpea, barley, and lentil through pre-breeding and genetic enhancement. At ICRISAT, pre-breeding activities in chickpea, groundnut, and pigeonpea are in progress using promising exotic landraces and wild species as donors along with popular varieties as recipients. The aim is to develop new gene pools with a higher frequency of useful genes, wider adaptability, and a broader genetic base for important agronomic and nutrition-related traits, as well as for resistance/tolerance to important biotic/abiotic stresses. Emphasis is being given to introgress genes for resistance against *Ascochyta* blight, *Botrytis*gray mold, and *Helicoverpa* pod borer in chickpea; for sterility mosaic disease, *Phytophthora* blight, and pod borer in pigeonpea; and for LLS, peanut stem necrosis disease (PSND), and aflatoxin in groundnut. Among the three SAT legumes, cultivated groundnut has the major need for pre-breeding activities due to a narrow gene pool. The tetraploidization event occurred during the evolution of cultivated groundnut has restricted the exchange of genomic regions between wild relatives and cultivated groups. Fortunately, wide crosses resulted in the development of highly diverse autotetraploids and allotetraploids. Simpson et al. (1993) and Fávero et al. (2006) reported development of three amphiploids using a range of wild AA- and BB- genome species such as *A. cardenasii, A. diogoi,* and *A. batizocoi, A. ipaënsis, A. duranensis, A. gregoryi,* and *A. linearifolium*. In order to diversify the primary gene pool, new amphiploid and autotetraploid groundnuts have been developed at ICRISAT from diploid hybrids such as *A. magna* × *A. valida, A. magna* × *A. batizocoi, A. batizocoi* × *A. cardenasii, A. batizocoi* × *A. duranensis, A. ipaënsis* × *A. duranensis, A. valida* × *A. duranensis, A. duranensis* × *A. cardenasii, A. kempff mercadoi* × *A. stenosperma, A. diogoi* × *A. cardenasii, A. duranensis* × *A. batizocoi, A. cardenasii* × *A. diogoi, A. duranensis* × *A. valida, A. duranensis* × *A. ipaënsis, A. kempff mercadoi* × *A. hoehnei, A. duranensis* × *A. batizocoi,* and *A. diogoi*× *A. cardenasii* (Mallikarjuna et al., 2011c). Evaluation of a few of these amphiploids has shown resistance to LLS (Mallikarjuna et al., 2012) and PBND (Shilpa et al., 2013). Using wild species of chickpea and pigeonpea, and synthetic amphiploids in groundnut, GWI lines and advanced-backcross (AB) populations are being developed in the genetic background of elite varieties following marker-assisted selection. (For developing AB-population, a selected wild relative is crossed to an elite cultivar followed by two to three backcrosses with targeted elite cultivar to develop segregating BC_2F_2 or BC_2F_3 populations). Phenotyping and genotyping of these (GWI and AB) populations should help in identifying the lines with enhanced genetic base and minimum linkage drag for use in future breeding programs, as well as to find out the markers associated with traits of interest.

INITIATIVES AND HOPE FOR ENRICHING CULTIVATED GENE POOL THROUGH GENOMICS-ASSISTED PRE-BREEDING

The availability of genomic resources has provided a solution to monitor allele/genome-specific introgression in GWI lines, along with identification of marker-trait association using AB populations. Genomics tools provide efficient tracking for desired and non-desired alien alleles among breeding lines using AB-QTL based breeding approach. Thus, integration of genomic tools with conventional breeding approaches promises to enrich the cultivated gene pool through genomics-assisted pre-breeding. These efforts will help in harnessing the rich diversity of wild relatives possessing superior alleles lost either during domestication or breeding, as well as for exploiting novel alleles. In AB-QTL based breeding approach, genotypic and phenotypic data generated on AB populations are used for identification of QTLs for several economically important agronomic traits. In addition, it also provides an opportunity to select for QTL-near isogenic lines (NILs) in a short time span along with the development of introgression lines with desired features for use in hybridization programmes.

As mentioned in previous section, attempts are being made at ICRISAT to introgress stress resistance through AB-breeding approach in grain legumes. In pigeonpea, ICRISAT is in the process of developing two backcross populations (ICPL 87119 × ICPW 29 and ICPL 87119 × ICPW 12) for conducting AB-QTL analysis and their subsequent use in AB-breeding. Among wild relatives used, ICPW 29 belongs to *C. cajanifolius* while ICPW 12 to *C. acutifolius* species. From the above crosses, BC_2F_3 seeds have been generated for multilocation phenotyping and selection. In groundnut, using two of the above mentioned synthetics, two AB-QTL mapping populations from the crosses between ICGV 91114 (cultivated) and ISATGR 1212 (*A. duranensis* ICG 8123 × *A. ipaënsis* ICG 8206, synthetic amphiploid); and between ICGV 87846 (cultivated) and ISATGR 265-5A (*A. kempff-mercadoi* ICG 8164 × *A. hoehnei* ICG 8190, synthetic amphiploid) have been developed at ICRISAT. Considerable variability has been observed in these two populations for pod and seed traits (Figure 3). A subset of these two populations has been phenotyped and genotyped with DArT markers to construct genetic maps and conducting AB-QTL analysis. The backcross (BC_1F_1 and BC_2F_1) lines carrying the alien genomic segments with maximum recurrent parent genomic regions provided optimal distribution of the synthetic genome introgressions (Foncéka et al., 2009).

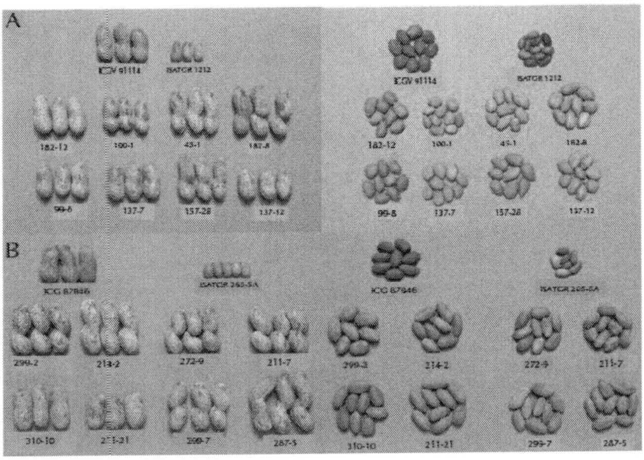

Figure 3. Variability for pod and seed traits in AB-QTL populations derived from using synthetics in crosses between (A) ICGV 91114 × ISATGR 1212 (above) and (B) ICGV 87846 × ISATGR 265-5A (below).

In case of chickpea multi-parent advanced generation intercross (MAGIC) populations are being developed to enhance the genetic base. Eight elite lines/cultivars (ICC 4958, ICCV 10, JAKI 9218, JG 11, JG 130, JG 16, ICCV 97105, and ICCV 00108) were selected from Ethiopia, Kenya and India for development of a MAGIC population for *desi* chickpea. Twenty-eight two-way, 14 four-way and seven eight-way crosses were made to develop this MAGIC population. The best progenies from each generation are being used in various national breeding programs by the breeders. Currently, 1200 F_6 are in the field (crop season 2012-2013). A total of 44 pre-breeding populations have also been developed.

The availability of large number of markers and cost-effective genotyping systems in recent years has ensured that linkage drag has been limited through stringent background selection and monitoring of the wild genomic regions. Hence, keeping in view the low genetic diversity in all the three legume crops, deployment of genomic-assisted pre-breeding is needed for accelerated and efficient enrichment of the cultivated gene pool. This should lead to the incorporation of desired/favorable alleles from wild relatives, increasing genetic gain and ultimately improving yield under adverse conditions and diverse cultural practices.

CONCLUSIONS

For grain legume improvement, sufficient genetic diversity exists in the form of landraces and wild relatives, which carry several useful genes for cultivar improvement. However, utilization of these resources in breeding programs is time-consuming and resource-demanding. To overcome this, pre-breeding activities should be initiated to generate new genetic variability using promising landraces and wild relatives for use by the breeders in crop improvement programs. Pre-breeding should focus on the continuous supply of useful variability into the breeding pipeline to develop new high-yielding cultivars with a broad genetic base; pre-breeding should not focus on increasing yield. Though pre-breeding is useful to enrich the primary gene pool for cultivar improvement, it is a time-consuming and difficult affair as well. As mentioned by Nass and Paterniani (2000), promising results can only be obtained in 5–10 years depending upon the information available in the beginning of the program. Further, linkage drag associated with utilizing wild relatives makes the pre-breeding activities much more cumbersome. Genomic-assisted pre-breeding will help to overcome the linkage drag and will facilitate focused transfer of useful genes/segments from wild relatives for genetic enhancement of grain legumes.

ACKNOWLEDGMENTS

Authors thank Dr. Nalini Mallikarjuna for sharing the material and Dr. Manish Pandey, Dr. Mahender Thudi, and Dr. Rachit Saxena for their suggestions in finalizing this manuscript.

REFERENCES

1. Acosta-Gallegos J., Kelly J. D., Gepts P. (2007). Prebreeding in common bean and use of genetic diversity from wild germplasm. Crop Sci. 47, S44–S59 10.2135/cropsci2007.04.0008IPBS
2. Ahmed K. (2008). Evaluation of Groundnut (Arachis hypogaea L.) Germplasm (Mini core collection). M.Sc. Thesis, University of Agricultural Sciences, Dharwad; Bangalore.
3. Ariyanayagam R. P., Nageshwara A., Zaveri P. P. (1995). Cytoplasmic genic male sterility in interspecific matings of pigeonpea. Crop Sci. 35, 981–985 10.2135/cropsci1995.0011183X003500040008x
4. Badami P. S., Mallikarjuna N., Moss J. P. (1997). Interspecific hybridization between *Cicer arietinum* and C. Pinnatifidum. Plant Breed. 116, 393–395 10.1111/j.1439-0523.1997.tb01019.x
5. Biabani A., Carpenter-Boggs L., Coyne C. J., Taylor L., Smith J. L., Higgins S. (2011).Nitrogen fixation potential in global chickpea mini core collection. Biol. Fertil. Soils 47, 679–685 10.1007/s00374-011-0574-0
6. Cantrell R. P., Christopher R., Dowswell C. R., Paliwal R. L. (1996). Maize in the Third World. Boulder, CO: Westview Press
7. Chaturvedi S. K., Nadarajan N. (2010). Genetic enhancement for grain yield in chickpea – accomplishments and resetting research agenda. Electron. J. Plant Breed. 1, 611–615
8. Dalrymple D. G. (1986). Development and Spread of High-Yielding Wheat Varieties in Developing Countries. 7th Edn Washington, DC: US Agency for International Development
9. Damicone J. P., Holbrook C. C., Jr., Smith D. L., Melouk H. A., Chamberlin K. D. (2010).Reaction of the core collection of peanut germplasm to Sclerotinia blight and pepper spot.Peanut Sci. 37, 1–11 10.3146/PS09-001.1
10. Dubey A., Farmer A., Schlueter J., Cannon S. B., Abernathy B., Tuteja R., et al. (2011).Defining the transcriptome assembly and its use for genome dynamics and transcriptome profiling studies in pigeonpea (*Cajanus cajan* L.). DNA Res. 18, 153–164 10.1093/dnares/dsr007

11. Dwivedi S. L., Stalker H. T., Blair M. W., Bertioli D. J., Upadhyaya H. D., Nielen S., et al. (2008a). Enhancing crop gene pools with beneficial traits using wild relatives. Plant Breed. Rev. 30, 179–230 10.1002/9780470380130.ch3

12. Dwivedi S. L., Puppala N., Upadhyaya H. D., Manivannan N., Singh S. (2008b). Developing a core collection of peanut specific to Valencia market type. Crop Sci. 48, 625–632 10.2135/cropsci2007.04.0240

13. FAO. (2010). The Second Report on the State of the World's Plant Genetic Resources for Food and Agriculture. Commission on Genetic Resources for Food and Agriculture (CGRFA). Rome: Food and Agriculture Organization of the United Nations (FAO)

14. Fávero A. P., Simpson C. E., Valls J. F. M., Vello N. A. (2006). Study of the evolution of cultivated peanut through crossability studies among *Arachis ipaënsis*, *A.* duranensis, and *A. hypogaea*. Crop Sci. 46, 1546–1552

15. Foncéka D., Hodo-Abalo T., Rivallan R., Faye I., Sall M. N., Ndoye O., et al. (2009). Genetic mapping of wild introgressions into cultivated peanut: a way toward enlarging the genetic basis of a recent allotetraploid.BMC Plant Biol. 9:103 10.1186/1471-2229-9-103

16. FrankeM. D., BennemenT. B., HolbrookC. C. (1999). Identification of resistance to Rhizoctonia limb rot in a core collection of peanut germplasm. Plant Dis. 83, 944–948 10.1094/PDIS.1999.83.10.944

17. Frankel O. H., Brown A. H. D. (1984). Plant genetic resources today: a critical appraisal, inCrop Genetic Resources: Conservation and Evaluation, eds Holden J. H. W., Williams J. T., editors. (London: George Allen and Unwin LTD;), 249–257

18. Gaur P. M., Mallikarjuna N., Knights T., Beebe S., Debouck D., Mejia A., et al. (2009). Gene introgression in grain legumes, in International Conference on Grain Legumes: Quality Improvement, Value Addition and Trade (Kanpur:).

19. Gowda C. L. L., Upadhyaya H. D., Dronavalli N., Singh S. (2011). Identification of large-seeded high-yielding stable kabuli chickpea germplasm lines for use in crop improvement.Crop Sci. 51, 198–209 10.2135/cropsci2010.01.0078

20. Graham P. H., Vance C. P. (2003). Legumes: importance and constraints to greater use. Plant Physiol. 131, 872–877 10.1104/pp.017004
21. Hamidou F., Ratnakumar P., Halilou O., Mponda O., Kapewa T., Monyo E., et al. (2011).Selection of intermittent drought tolerant lines across years and locations in the reference collection of groundnut (*Arachis hypogaea* L.). Field Crops Res. 126, 189–199 10.1016/j.fcr.2011.10.009
22. Hammons R. O. (1970). Registration of Spancross peanuts (Reg. No. 3). Crop Sci. 10, 459–460
23. Hannan R. M., Kaiser W. J., Muehlbauer F. J. (1994). Development and utilization of the USDA chickpea germplasm core collection, in Agronomy Abstracts 1994, (Madison, WI: ASA;), 217.
24. Hiremath P. J., Farmer A., Cannon S. B., Woodward J., Kudapa H., Tuteja R., et al. (2011).Large-scale transcriptome analysis in chickpea (*Cicer arietinum* L.), an orphan legume crop of the semi-arid tropics of Asia and Africa. Plant Biotechnol. J. 9, 922–931 10.1111/j.1467-7652.2011.00625.x
25. Hiremath P. J., Kumar A., Penmetsa R. V., Farmer A., Schlueter J. A., Chamarthi S. K., et al. (2012). Large-scale development of cost-effective SNP marker assays for diversity assessment and genetic mapping in chickpea and comparative mapping in legumes. Plant Biotechnol. J.10, 716–732 10.1111/j.1467-7652.2012.00710.x
26. Holbrook C. C., Anderson W. F., Pittman R. N. (1993). Selection of a core collection from the U.S. Germplasm collection of peanut. Crop Sci. 33, 859–861 10.2135/cropsci1993.0011183X003300040044x
27. Holbrook C. C., Dong W. (2005). Development and evaluation of a mini core collection for the U.S. Peanut Germplasm Collection. Crop Sci. 45, 1540–1544 10.2135/cropsci2004.0368
28. ICARDA. (1995). Annual report for 1995, in Germplasm Program Legumes (Aleppo:), 210.
29. Jadhav D. R., Mallikarjuna N., Sharma H. C., Saxena K. B. (2012). Introgression of*Helicoverpa armigera* resistance from *Cajanus acutifolius* – a wild relative from secondary gene pool of pigeonpea (*Cajanus cajan*). Asian. J. Agric. Sci. 4, 242–248

30. Jiang H. F., Ren X. P., Liao B. S., Huang J. Q., Lei Y., Chen B. Y., et al. (2008). Peanut core collection established in china and compared with ICRISAT mini core collection. Acta Agron. Sin. 34, 25–30 10.3724/SP.J.1006.2008.00025

31. Jiang H. F., Ren X. P., Wang S. Y., Zhang X. J., Huang J. Q., Liao B. S., et al. (2010).Development and evaluation of peanut germplasm with resistance to *Aspergillu flavus* from core collection. Acta Agron. Sin. 36, 428–434 10.3724/SP.J.1006.2010.00428

32. Kashiwagi J., Krishnamurthy L., Singh S., Upadhyaya H. D. (2006). Variation of SPAD chlorophyll meter readings (SCMR) in the mini core germplasm collection of chickpea. J. SAT Agric. Res. 2, 1–3

33. Kashiwagi J., Krishnamurthy L., Upadhyaya H. D., Gaur P. M. (2008). Rapid screening technique for canopy temperature status and its relevance to drought tolerance improvement in chickpea. J. SAT Agric. Res. 6, 105–104

34. Kashiwagi J., Krishnamurthy L., Upadhyaya H. D., Krishna H., Chandra S., Vadez V., et al. (2005). Genetic variability of drought-avoidance root traits in the mini core germplasm collection of chickpea (*Cicer arietinum* L.). Euphytica 146, 213–222 10.1007/s10681-005-9007-1

35. Kashiwagi J., Upadhyaya H. D., Krishnamurthy L. (2010). Significance and genetic diversity of SPAD chlorophyll meter reading in chickpea germplasm in the semi-arid environments. J. Food Legumes 23, 99–105

36. Kashiwagi J., Upadhyaya H. D., Krishnamurthy L., Singh S. (2007). Identification of high-yielding large-seeded kabuli chickpeas with drought avoidance root traits. J. SAT Agric. Res.3, 1–2

37. Kaul J., Kumar S., Gurha S. N. (2007). Evaluation of exotic germplasm of kabuli chickpea.Indian J. Plant Genet. Resour. 20, 160–164

38. Khera P., Sriswathi M., Roorkiwal M., Pandey M., Upadhyaya H. D., Janila P., et al. (2013).Cost-effective KASPar assays for SNP genotyping and diversity analysis in the reference set of groundnut, in Plant and Animal Genome XXI Conference. Plant and Animal Genome, (San Diego, CA:), P0363.

39. Kibret K. T. (2012). Development and utilization of genetic diversity based Ethiopian chickpea (*Cicer arietinum* L.) germplasm core collection for association mapping. *Ph.D. Thesis, International Crops Research Institute for the Semi-Arid Tropics (ICRISAT),Hyderabad*

40. Krishnamurthy L., Gaur P. M., Basu P. S., Chaturvedi S. K., Tripathi S., Vadez V., et al. (2011a). Large genetic variation for heat tolerance in the reference collection of chickpea (*Cicer arietinum* L.) germplasm. Plant Genet. Resour. 9, 59–69 10.1017/S1479262110000407

41. Krishnamurthy L., Upadhyaya H. D., Saxena K. B., Vadez V. (2011b). Variation for temporary water logging response within the mini core pigeonpea germplasm. J. Agric. Sci.10, 1–8

42. Krishnamurthy L., Turner N. C., Gaur P. M., Upadhyaya H. D., Varshney R. K., Siddique K. H. M., et al. (2011c). Consistent variation across soil types in salinity resistance of a diverse range of chickpea (*Cicer arietinum* L.) genotypes. J. Agron. Crop Sci. 197, 214–227 10.1111/j.1439-037X.2010.00456.x

43. Krishnamurthy L., Kashiwagi J., Gaur P. M., Upadhyaya H. D., Vadez V. (2010). Sources of tolerance to terminal drought in the chickpea (*Cicer arietinum L.*) mini core germplasm. Field Crops Res. 119, 322–330

44. Kumar J., van Rheenen H. A. (2000). A major gene for time of flowering in chickpea. J. Hered. 91, 67–68 10.1093/jhered/91.1.67

45. Kumar S., Gupta S., Chandra S., Singh B. B. (2004). How wide is genetic base of pulse crops?" in Pulses in new perspective, in Proceedings of the National Symposium on Crop Diversification and Natural Resources Management, eds Ali M., Singh B. B., Shiv Kumar, Vishwa Dhar, editors. (Kanpur: Indian Society of Pulses Research and Development, Indian Institute of Pulses Research;), 211–221

46. Kusuma V. P., Yugandhar G., Ajay B. C., Gowda M. V. C., Upadhyaya H. D. (2007).Identification of sources of multiple disease resistance in groundnut (*Arachis hypogaea* L.) mini core, in ISOR National Seminar (Hyderabad: Directorate of Oilseeds Research;), 31–32

47. Lulsdorf M., Mallikarjuna N., Clarke H., Tar'An B. (2005). Finding solutions for interspecific hybridization problems in chickpea (*Cicer arietinum* L.), in 4th International Food Legumes Research Conference (New Delhi:).
48. Malhotra R. S., Singh K. B., Vito M., Greco N., Saxena M. C. (2002). Registration of ILC 10765 and ILC 10766 chickpea germplasm lines resistant to cyst nematode. Crop Sci. 42, 1756 10.2135/cropsci2002.1756
49. Mallikarjuna N. (1999). Ovule and embryo culture to obtain hybrids from interspecific incompatible pollinations in chickpea. Euphytica 110, 1–6 10.1023/A:1003621908663
50. Mallikarjuna N., Jadhav D., Reddy M. V., Dutta-Tawar U. (2005). Introgression of*Phytophthora* blight disease resistance from *Cajanus platycarpus* into short duration pigeonpea *Cajanus cajan* (L.) Millsp.. Indian J. Genet. 65, 261–263
51. Mallikarjuna N., Jadhav D. R. (2008). Techniques to produce hybrids between *Cicer arietinum* L. x Cicer pinnatifidum Jaub. Indian J. Genet. 68, 398–405
52. Mallikarjuna N., Jadhav D. R., Reddy K., Husain F., Das K. (2012). Screening new *Arachis*amphidiploids, autotetraploids for resistance to late leaf spot by detached leaf technique. Eur. J. Plant Pathol. 132, 17–21 10.1007/s10658-011-9859-2
53. Mallikarjuna N., Jadhav D. R., Srikant S., Saxena K. B. (2011a). *Cajanus platycarpus*(Benth.) Maesen as the donor of new pigeonpea cytoplasmic male sterile (CMS) system.Euphytica 182, 65–71 10.1007/s10681-011-0488-9
54. Mallikarjuna N., Senapathy S., Jadhav D. R., Saxena K. B., Sharma H. C., Upadhyaya H. D., et al. (2011b). Progress in the utilization of *Cajanus platycarpus* (Benth.) Maesen in pigeonpea improvement. Plant Breed. 130, 507–514 10.1111/j.1439-0523.2011.01870.x
55. MallikarjunaN.,SenthilvelS.,HoisingtonD.(2011c).Development of new sources of tetraploid Arachis to broaden the genetic base of cultivated groundnut (*Arachis hypogaea* L.).Genet. Resour. Crop Evol. 58, 889–907 10.1007/s10722-010-9627-8
56. Mallikarjuna N., Sharma H. C., Upadhyaya H. D. (2007). Exploitation of wild relatives of pigeonpea and chickpea for resistance to *Helicoverpa armigera*. SAT eJ. 3, 4

57. Marshall D. R. (1989). Limitations to the use of germplasm collections, in The Use of Plant Genetic Resources, eds Brown A. D. H., Frankel O. H., Marshall D. R., Williams J. T., editors. (Cambridge: University Press;), 105–120

58. Meena H. P., Kumar J., Upadhyaya H. D., Bharadwaj C., Chauhan S. K., Verma A. K., et al. (2010). Chickpea mini core germplasm collection as rich sources of diversity for crop improvement. J. SAT Agric. Res. 8, 1–5

59. Mikel M. A., Diers B. W., Nelson R. L., Smith H. H. (2010). Genetic diversity and agronomic improvement of North American soybean germplasm. Crop Sci. 50, 1219–1229 10.2135/cropsci2009.08.0456

60. Moss J. P., Singh A. K., Nigam S. N., Hilderbrand G. L., Goviden N., Ismael F. M. (1998).Registration of ICGV-SM 87165 peanut germplasm. Crop Sci. 38, 572 10.2135/cropsci1998.0011183X003800020099x

61. Nagy E. D., Guo Y., Tang S., Bowers J. E., Okashah R. A., Taylor C. A., et al. (2012). A high-density genetic map of *Arachis duranensis*, a diploid ancestor of cultivated peanut. BMC Genomics 13:469 10.1186/1471-2164-13-469

62. Nass L. L., Paterniani E. (2000). Pre-breeding: a link between genetic resources and maize breeding. Sci. Agric. 57, 581–587 10.1590/S0103-90162000000300035

63. Nigam S. N., Hildebrand G. H., Bock K. R., Ismael F. M., Govinden N., Subrahmanyam P., et al. (1998). Registration of ICGV-SM 85048 peanut germplasm. Crop Sci. 38, 572–573 10.2135/cropsci1998.0011183X003800020100x

64. Pande S., Desai S., Sharma M. (2010). Impact of climate change on rainfed crop diseases: current status and future research needs, in National Symposium on Climate Change and Rainfed Agriculture, eds Venkateswaru B., Mishara P. K., Rao V. U. M., Ravindra Chary G., Srinivasa Rao C. H., Sharma K. L., K. A., et al., editors. (Hyderabad: Society of Dryland Agriculture, Central Research Institute for Dryland Agriculture;), 55–59

65. Pande S., Kishore G. K., Upadhyaya H. D., Rao J. N. (2006). Identification of multiple diseases resistance in mini core collection of chickpea. Plant Dis. 90, 1214–1218 10.1094/PD-90-1214

66. Pande S., Sharma M. (2011). Climate change and changing scenario of plant diseases in semi-arid tropics, in 64th Annual Meeting of the Society and National Symposium on Biology of Infection, Immunity and Disease Control in Pathogen-Plant Interactions (Hyderabad: University of Hyderabad;).

67. Parameshwarappa S. G., Salimath P. M. (2008). Field screening of chickpea genotypes for drought resistance. Karnataka J. Agric. Sci. 21, 113–114

68. Parameshwarappa S. G., Salimath P. M., Upadhyaya H. D., Patil S. S., Kajjidoni S. T. (2011a). Genetic divergence under three environments in a mini core collection of chickpea (*Cicer arietinum* L.). Indian J. Plant Genet. Resour. 24, 177–185

69. Parameshwarappa S. G., Salimath P. M., Upadhyaya H. D., Patil S. S., Kajjidoni S. T. (2011b). Genetic variability studies in mini core collections of chickpea (*Cicer arietinum* L.) under different environments. Indian J. Plant Genet. Resour. 24, 43–48

70. Parameshwarappa S. G., Salimath P. M., Upadhyaya H. D., Patil S. S., Kajjidoni S. T., Patil B. C. (2010). Characterization of drought tolerant accessions identified from the mini core of chickpea (*Cicer arietinum* L.). Indian J. Genet. Plant Breed. 70, 125–131

71. Pouresmael M., Akbari M., Vaezi S., Shahmoradi S. (2009). Effects of drought stress gradient on agronomic traits in Kabuli chickpea core collection. Iran. J. Crop Sci. 11, 308–324

72. Ramgopal D., Srivastava R. K., Pande S., Rathore A., Jadhav D. R., Sharma M., et al. (2012).Introgression of Botrytis gray mold resistance gene from *Cicer reticulatum* (*bgmr1cr*) and *C. echinospermum* (*bgmr1ce*) to chickpea. Plant Genet. Resour. 1–6.

73. Rao N. K., Reddy L. J., Bramel P. J. (2003). Potential of wild species for genetic enhancement of some semi-arid food crops. Genet. Resour. Crop Evol. 50, 707–721 10.1023/A:1025055018954

74. Reddy L. J., Nigam S. N., Moss J. P., Singh A. K., Subrahmanyam P., Mc Donald D., et al. (1996). Registration of ICGV 86699 peanut germplasm line with multiple disease and insect resistance. Crop Sci. 36, 821 10.2135/cropsci1996.0011183X003600030072x

75. Reddy L. J., Saxena K. B., Jain K. C., Singh U., Green J. M., Sharma D., et al. (1997).Registration of high-protein pigeonpea

elite germplasm ICPL 87162. Crop Sci. 37, 294 10.2135/cropsci 1997.0011183X003700010066x

76. Reddy L. J., Upadhyaya H. D., Gowda C. L. L., Singh S. (2005). Development of core collection in pigeonpea *Cajanus cajan* (L.) Millsp. using geographic and qualitative morphological descriptors. Genet. Resour. Crop Evol. 52, 1049–1056 10.1007/s10722-004-6152-7

77. Rockland L. B., Radke T. M. (1981). Legume protein quality. Food Technol. 35, 79

78. Saxena K. B., Faris D. G., Kumar R. V. (1987). Relationship between seed size and protein content in newly developed high protein lines of pigeonpea. Plant Foods Hum. Nutr. 36, 335–340 10.1007/BF01892354

79. Saxena K. B., Kumar R. V. (2003). Development of cytoplasmic nuclear male-sterility system in pigeonpea using *C. scarabaeoides* (L.) Thours. Indian J. Genet. 63, 225–229

80. Saxena K. B., Kumar R. V., Dalvi V. A., Mallikarjuna N., Gowda C. L. L., Singh B. B., et al. (2005). Hybrid breeding in grain legumes–a success story of pigeonpea, in Proceedings of the International Food Legumes Research Conference, eds Khairwal M. C., Jain H. K., editors. (New Delhi:).

81. Saxena K. B., Sultana R., Mallikarjuna N., Saxena R. K., Kumar R. V., Sawargaonkar S. L., et al. (2010). Male-sterility systems in pigeonpea and their role in enhancing yield. Plant Breed.129, 125–134 10.1111/j.1439-0523.2009.01752.x

82. Saxena R. K., Varma Penmetsa R., Upadhyaya H. D., Kumar A., Carrasquilla-Garcia N., Schlueter J. A., et al. (2012). Large-scale development of cost-effective single-nucleotide polymorphism marker assays for genetic mapping in pigeonpea and comparative mapping in legumes. DNA Res. 19, 449–461 10.1093/dnares/dss025

83. Serraj R., Krishnamurthy L., Upadhyaya H. D. (2004). Screening chickpea mini core germplasm for tolerance to soil salinity. Int. Chickpea Pigeonpea Newsl. 11, 29–32

84. Sharma H. C., Sujana G., Rao D. M. (2009). Morphological and chemical components of resistance to pod borer, *Helicoverpa armigera* in wild relatives of pigeonpea. Springer, SBM, Dordrecht,

Netherlands. Arthropod Plant Interact. 3, 151–161 10.1007/s11829-009-9068-5

85. Sharma M., Pande S., Pathak M., Narayana Rao J., Anil Kumar P., Madhusudun Reddy D., et al. (2006). Prevalence of *Phytophthora* blight of pigeonpea in the Deccan Plateau. J. Plant Pathol. 22, 309–313 10.5423/PPJ.2006.22.4.309

86. Sharma M., Rathore A., Mangala U. N., Ghosh R., Upadhyaya H. D., Sharma S., et al. (2012).New sources of resistance to *Fusarium* wilt and sterility mosaic disease in a mini core collection of pigeonpea germplasm. Eur. J. Plant Pathol. 133, 707–714 10.1007/s10658-012-9949-9

87. Sharma S., Upadhyaya H. D., Roorkiwal M., Varshney R. K., Gowda C. L. L. (2013).Chickpea, in Genetic and Genomic Resources of Grain Legume Improvement, eds Singh M., Upadhyaya H. D., Bisht I. S., editors. (London; Waltham, MA: Elsevier Inc;), 81–104

88. Shilpa K., Sunkad G., Kurella S., Marri S., Padmashree K., Jadhav D. R., et al. (2013).Biochemical composition and disease resistance in newly synthesized amphidiploid and autotetraploid peanuts. Food and Nutr. Sci. 4, 169–176 10.4236/fns.2013.42024

89. Simpson C. E., Nelson S. C., Starr J., Woodward K. E., Smith O. D. (1993). Registration of TxAG-6 and TxAG-7 peanut germplasm lines. Crop Sci. 33, 1418 10.2135/cropsci1993.0011183X003300060079x

90. Simpson C. E., Smith O. D. (1975). Registration of Tamnut 74 peanut (Reg. No. 19). Crop Sci. 15, 603–604 10.2135/cropsci1975.0011183X001500040050x

91. Simpson C. E., Starr J. L. (2001). Registration of COAN peanut. Crop Sci. 41, 918 10.2135/cropsci2001.413918x

92. Simpson C. E., Starr J. L., Church G. T., Burow M. D., Paterson A. H. (2003). Registration of 'Nema TAM' peanut. Crop Sci. 43, 1561 10.2135/cropsci2003.1561

93. Simpson C. E., Starr J. L., Nelson S. C., Woodard K. E., Smith O. D. (1993). Registration of TxAG-6 and TxAG-7 peanut germplasm. Crop Sci. 33, 1418 10.2135/cropsci1993.0011183X003300060079x

94. Singh A. K., Dwivedi S. L., Pande S., Moss J. P., Nigam S. N., Sastri D. C. (2003).Registration of rust and late leaf spot resistant peanut germplasm lines. Crop Sci. 43, 440–441 10.2135/cropsci2003.0440
95. Singh K. B., Ocampo B. (1997). Exploitation of wild *Cicer* species for yield improvement in chickpea. Theor. Appl. Genet. 95, 418–423 10.1007/s001220050578
96. Singh N. P., Singh A., Asthana A. N., Singh A. (1999). Studies on inter-specific crossability barriers in chickpea. Ind. J. Pulses Res. 12, 13–19
97. Singh R. P., Singh B. D. (1989). Recovery of rare interspecific hybrids of gram *Cicer arietinum* x *C. cuneatum* L. through tissue culture. Curr. Sci. 58, 874–876
98. Singh S., Gumber R. K., Joshi N., Singh K. (2005). Introgression from wild *Cicer reticulatum*to cultivated chickpea for productivity and disease resistance. Plant Breed. 124, 477–480 10.1111/j.1439-0523.2005.01146.x
99. Srivastava N. (2010). Identification of Contrasting Parental Material for Salinity Tolerance Based on Morphological, Physiological, and Biochemical Study and Molecular Diversity by SSR Molecular Markers in Groundnut (Arachis hypogaea). Ph.D., Thesis, Jawaharlal Nehru Technological University, Hyderabad.
100. Srivastava N., Vadez V., Upadhyaya H. D., Saxena K. B. (2006). Screening for intra and interspecific variability for salinity tolerance in pigeonpea (*Cajanus cajan*) and its wild species. J. SAT Agric. Res. 2, 12
101. Subbarao G. V., Johansen C., Kumar Rao J. V. D. K., Jana M. K. (1990). Salinity tolerance in F1 hybrids of pigeonpea and a tolerant wild relative. Crop Sci. 30, 785–788 10.2135/cropsci1990.0011183X003000040005x
102. Sujana G., Sharma H. C., Rao D. M. (2008). Antixenosis and antibiosis components of resistance to pod borer *Helicoverpa armigera* in wild relatives of pigeonpea. Int. J. Trop. Insect Sci. 28, 191–200
103. Tar'an B., Warkentin T. D., Vandenberg A., Holm F. A. (2010). Variation in chickpea germplasm for tolerance to imazethapyr and imazamox herbicides. Can. J. Plant Sci. 90, 139–142

104. Thudi M., Bohra A., Nayak S. N., Varghese N., Shah T. M., Penmetsa R. V., et al. (2011).Novel SSR markers from BAC-end sequences, DArT arrays and a comprehensive genetic map with 1, 291 marker loci for chickpea (*Cicer arietinum* L.). PLoS ONE 6:e27275 10.1371/journal.pone.0027275

105. Tikka S. B. S., Parmar L. D., Chauhan R. M. (1997). First record of cytoplasmic-genic male sterility system in pigeonpea (*Cajanus cajan* (L.) Millsp.) through wide hybridization. Gujarat Agric. Univ. Res. J. 22, 160–162

106. Upadhyaya H. D. (2005). Variability for drought resistance related traits in the mini core collection of peanut. Crop Sci. 45, 1432–1440 10.2135/cropsci2004.0389

107. Upadhyaya H. D. (2008). Crop Germplasm and wild relatives: a source of novel variation for crop improvement. Korean J. Crop Sci. 53, 12–17

108. Upadhyaya H. D., Ortiz R. (2001). A mini core subset for capturing diversity and promoting utilization of chickpea genetic resources in crop improvement. Theor. Appl. Genet. 102, 1292–1298 10.1007/s00122-001-0556-y

109. Upadhyaya H. D., Bramel P. J., Ortiz R., Singh S. (2002). Developing a mini core of peanut for utilization of genetic resources. Crop Sci. 42, 2150–2156 10.2135/cropsci2002.2150

110. Upadhyaya H. D., Dronavalli N., Gowda C. L. L., Singh S. (2011a). Identification and evaluation of chickpea germplasm for tolerance to heat stress. Crop Sci. 51, 2079–2094 10.2135/cropsci2011.01.0018

111. Upadhyaya H. D., Thudi M., Dronavalli N., Gujaria N., Singh S., Sharma S., et al. (2011b).Genomic tools and germplasm diversity for chickpea improvement. Plant Genet. Resour. 9, 45–48 10.1017/S1479262110000468

112. Upadhyaya H. D., Salimath P. M., Gowda C. L. L., Singh S. (2007a). New early-maturing germplasm lines for utilization in chickpea improvement. Euphytica 157, 195–208 10.1007/s10681-007-9411-9

113. Upadhyaya H. D., Dwivedi S. L., Gowda C. L. L., Singh S. (2007b). Identification of diverse germplasm lines for agronomic traits in a chickpea (*Cicer arietinum* L.) core collection for use in

crop improvement. Field Crops Res. 100, 320–326 10.1016/j.fcr.2006.08.008

114. Upadhyaya H. D., Mallikarjuna Swamy B. P., Goudar P. V. K., Kullaiswamy B. Y., Singh S. (2005). Identification of diverse groundnut germplasm through multienvironment evaluation of a core collection for Asia. Field Crops Res. 93, 293–299 10.1016/j.fcr.2004.10.007

115. Upadhyaya H. D., Sharma S., Dwivedi S. L. (2012a). Arachis, in Wild Crop Relatives: Genomic and Breeding Resources, Legume Crops and Forages, ed Kole C., editor. (Berlin; Heidelberg: Springer-Verlag Publisher;), 1–19

116. Upadhyaya H. D., Mukri G., Nadaf H. L., Singh S. (2012b). Variability and stability analysis for nutritional traits in the mini core collection of peanut. Crop Sci. 52, 168–178 10.2135/cropsci2011.05.0248

117. Upadhyaya H. D., Bramel P. J., Singh S. (2001a). Development of a chickpea core collection using geographic distribution and quantitative traits. Crop Sci. 41, 206–210 10.2135/cropsci2001.411206x

118. Upadhyaya H. D., Ortiz R., Bramel P. J., Singh S. (2001b). Development of Asia region groundnut core collection, in Paper Presented in the Diamond Jubilee Symposium on Hundred Years of Post-Mendelian Genetics - Retrospect and Prospects, Indian Agricultural Research Institute (New Delhi:), 43–44

119. Upadhyaya H. D., Ortiz R., Bramel P. J., Singh S. (2003). Development of a groundnut core collection using taxonomical, geographical and morphological descriptors. Genet. Resour. Crop Evol. 50, 139–148 10.1023/A:1022945715628

120. Upadhyaya H. D., Reddy L. J., Gowda C. L. L., Reddy K. N., Singh S. (2006a). Development of a mini core subset for enhanced and diversified utilization of pigeonpea germplasm resources. Crop Sci. 46, 2127–2132 10.2135/cropsci2006.01.0032

121. Upadhyaya H. D., Reddy L. J., Gowda C. L. L., Singh S. (2006b). Identification of diverse groundnut germplasm: sources of early-maturity in a core collection. Field Crops Res. 97, 261–267 10.1016/j.fcr.2005.10.010

122. Upadhyaya H. D., Reddy L. J., Gowda C. L. L., Singh S. (2009). Phenotypic diversity in cold-tolerant peanut (*Arachis hypogaea* L.) germplasm. Euphytica 165, 279–291 10.1007/s10681-008-9786-2

123. Upadhyaya H. D., Sharma S., Reddy K. N., Saxena R., Varshney R. K., Gowda C. L. L. (2013). Pigeonpea, in Genetic and Genomic Resources of Grain Legume Improvement, eds Singh M., Upadhyaya H. D., Bisht I. S., editors. (London; Waltham, MA: Elsevier Inc.), 181–198

124. Upadhyaya H. D., Yadav D., Dronavalli N., Gowda C. L. L., Singh S. (2010). Mini core germplasm collection for infusing genetic diversity in plant breeding programs. Electron. J. Plant Breed. 1, 1294

125. Vadez V., Krishnamurthy L., Serraj R., Gaur P. M., Upadhyaya H. D., Hoisington D. A., et al. (2007). Large variation in salinity tolerance in chickpea is explained by differences in sensitivity at reproductive stage. Field Crops Res. 104, 123–129 10.1016/j.fcr.2007.05.014

126. Valkoun J. J. (2001). Wheat pre-breeding using wild progenitors. Euphytica 119, 17–23 10.1023/A:1017562909881

127. Varshney R. K., Kudapa H., Roorkiwal M., Thudi M., Pandey M. K., Saxena R. K., et al. (2012a). Advances in genetics and molecular breeding of three legume crops of semi-arid tropics using next-generation sequencing and high-throughput genotyping technologies. J. Biosci. 37, 811–820 10.1007/s12038-012-9228-0

128. Varshney R. K., Chen W., Li Y., Bharti A. K., Saxena R. K., Schlueter J. A., et al. (2012b).Draft genome sequence of pigeonpea (*Cajanus cajan*), an orphan legume crop of resource-poor farmers. Nat. Biotechnol. 30, 83–89 10.1038/nbt.2022

129. Varshney R. K., Mohan S. M., Gaur P. M., Gangarao N. V. P. R., Pandey M. K., Bohra A., et al. (2013a). Achievements and prospects of genomics-assisted breeding in three legume crops of the semi-arid tropics. Biotechnol. Adv. Epub ahead of print. 10.1016/j.biotechadv.2013.01.001

130. Varshney R. K., Song C., Saxena R. K., Azam S., Yu S., Sharpe A. G., et al. (2013b). Draft genome sequence of chickpea

(*Cicer* arietinum) provides a resource for trait improvement.Nat. Biotechnol. 31, 240–246 10.1038/nbt.2491

131. Vellve R. (1992). Saving the Seeds: Genetic Diversity and European Agriculture. London: Earthscan Publication
132. Verma M. M., Ravi, Sandhu J. S. (1995). Characterization of the interspecific cross*Cicerarietinum,* L. x C. judaicum (Boiss.). Plant Breed. 114, 549–551 10.1111/j.1439-0523.1995.tb00856.x
133. Verma M. M., Sandhu J. S., Rrar H. S., Brar J. S. (1990). Crossability studies in different species of *Cicer*(L.). Crop Improv. 17, 179–181
134. Wanjari K. B., Patil A. N., Manapure P., Manjaya J. G., Manish P. (2001). Cytoplasmic male-sterility with cytoplasm from *Cajanus volubilis*. Ann. Plant Physiol. 13, 170–174.
135. Yadav S. S., Kumar J., Turner N. C., Berger J., Redden R., McNeil D., et al. (2004). Breading for improved productivity, multiple resistance and wide adaptation in chickpea (*Cicer arietinum* L.). Plant Genet. Resour. Charact. Util. 2, 181–187 10.1079/PGR200448

Chapter 10

Construction of a SNP-based Genetic Linkage map in Cultivated Peanut based on large scale Marker Development using next-Generation Double-Digest Restriction-site-Associated DNA Sequencing (ddRADseq)

Xiaojing Zhou, Youlin Xia, Xiaoping Ren, Yulin Chen, Li Huang, Shunmou Huang, Boshou Liao, Yong Lei, Liyin Yan, and Huifang Jiang

Key Laboratory of Biology and Genetic Improvement of Oil Crops, Ministry of Agriculture, Oil Crops Research Institute of the Chinese Academy of Agricultural Sciences, Wuhan, 430062 Hubei People's Republic of China

Nanchong Academy of Agricultural Sciences, Nanchong, 637000 Sichuan People's Republic of China

Databridge Technologies Corporation, Wuhan, 430062 Hubei People's Republic of China

Xiaojing Zhou,

ABSTRACT

Background

Cultivated peanut, or groundnut (*Arachis hypogaea* L.), is an important oilseed crop with an allotetraploid genome (AABB, $2n=4x=40$). In recent years, many efforts have been made to construct linkage maps in cultivated peanut, but almost all of these maps were constructed using low-throughput molecular markers, and most show a low density, directly influencing the value of their applications. With advances in next-generation sequencing (NGS) technology, the construction of high-density genetic maps has become more achievable in a cost-effective and rapid manner. The objective of this study was to establish a high-density single nucleotide polymorphism (SNP)-based genetic map for cultivated peanut by analyzing next-generation double-digest restriction-site-associated DNA sequencing (ddRADseq) reads.

Results

We constructed reduced representation libraries (RRLs) for two *A. hypogaea* lines and 166 of their recombinant inbred line (RIL) progenies using the ddRADseq technique. Approximately 175 gigabases of data containing 952,679,665 paired-end reads were obtained following Solexa sequencing. Mining this dataset, 53,257 SNPs were detected between the parents, of which 14,663 SNPs were also detected in the population, and 1,765 of the obtained polymorphic markers met the requirements for use in the construction of a genetic map. Among 50 randomly selected *in silico* SNPs, 47 were able to be successfully validated. One linkage map was constructed, which was comprised of 1,685 marker loci, including 1,621 SNPs and 64 simple sequence repeat (SSR) markers. The map displayed a distribution of the markers into 20 linkage groups (LGs A01–A10 and B01–B10), spanning a distance of 1,446.7 cM. The alignment of the LGs from this map was shown in comparison with a previously integrated consensus map from peanut.

Conclusions

This study showed that the ddRAD library combined with NGS allowed the rapid discovery of a large number of SNPs in the cultivated peanut. The first high density SNP-based linkage map for *A. hypogaea* was generated that can serve as a reference map for cultivated Arachis species and will be useful in genetic mapping. Our results contribute to the available molecular marker resources and to the assembly of a reference genome sequence for the peanut.

BACKGROUND

Cultivated peanut, or groundnut (*Arachis hypogaea* L.), is a major economic crop in most tropical and subtropical areas of the world and represents a significant source of oil and protein for human nutrition. Because this species is a self-pollinating allotetraploid (AABB, $2n = 4\times = 40$) with a large genome size (2800 Mb/1C) and a narrow genetic base, leading to very low DNA polymorphism, the development of molecular markers and genomic resources in peanut has always been a formidable task [1–3]. In recent years, many efforts have been made to construct linkage maps as the genetic basis for quantitative trait locus (QTL) analyses of important, complex traits. However, almost all the maps constructed using low-throughput molecular markers, e.g., restriction fragment length polymorphisms (RFLPs) and simple sequence repeats (SSRs), present a low density and are unable to provide precise information on the QTLs controlling the traits of interest [4–6]. In the tetraploid peanut, almost all of the existing linkage maps for single populations include fewer than 350 markers [5, 7], with the exception of two recently developed linkage maps that include over 1000 markers [8, 9]. In 2012, a single nucleotide polymorphism (SNP) marker-based genetic map was successfully constructed for the AA genome due to the greater simplicity of diploids [10], marking a step forward in the development of SNP markers for peanut. However, until recently, only sporadic SNP markers had been developed in cultivated peanuts, and no SNP marker-based genetic map has been previously reported.

SNPs are widely distributed in the genome and are the most abundant type of DNA variation currently used as a genetic marker [11]. Compared to markers based on size discrimination or hybridization, SNPs directly interrogate sequence variation and can reduce genotyping errors [12]. SNP discovery is amenable to high-throughput technology, such as next-generation sequencing (NGS) technologies, which produce DNA sequences at a rate several orders of magnitude faster than conventional methods, making them an excellent tool for use in genomics research.

The complexity of genomes can be overcome by using reduced representation libraries (RRLs), and the combination of RRLs with multiplex sequencing can improve the throughput of SNP identification and genotyping [13, 14]. RRLs are being used in a wide range of applications, including the construction of linkage maps, fine mapping of genes and association studies [15–17]. RRL was first and has usually been demonstrated through restriction site-associated DNA (RAD) tagging and NGS of RAD tags [18,19]. To increase the breadth of RADseq applications, the double-digest restriction-site-associated DNA sequencing (ddRADseq) method was developed by eliminating random shearing and explicitly using size selection to recover a tunable number of regions [20]. ddRADseq tags not only possess the advantages of RAD tags, such as allowing high-throughput, multiplexed sequencing and being amenable to genotyping, but they also provide improved efficiency and robustness compared to RAD. In *Brassica napus*, RRLs were constructed for two parents and 91 of their doubled haploid (DH) progenies using the ddRADseq technique, and restriction fragments in the size range of 141–420 bp were chosen to represent the reduced genome and to allow multiplex sequencing to be conducted [21]. SNPs were identified and genotyped from the high-quality polymorphism data, and a SNP bin map comprising 8,780 SNP loci, together with 47 SSR loci was constructed. Recknagel et al. [22] applied this technology to obtain a high-density linkage map for Cichlid fishes. A total of 755 markers were genotyped in 343 F_2 hybrids. The map resolved 25 linkage groups and spanned a total distance of

1,427 cM, with an average marker spacing distance of 1.95 cM [22]. These data suggest that ddRADseq technology can contribute to the construction of linkage maps through the identification and genotyping of SNPs across large numbers of individuals for a range of markers in both model and non-model species.

Through the utilization of NGS data, several bioinformatics approaches and tools have been developed for SNP discovery and genotyping in complex genomes. For instance, the GMAP alignment method and the Maq analysis method have been applied in soybean with stringent matching criteria (using only high-quality reads, unique mappings, multiple-reads SNP support) for high-throughput SNP discovery through RRL resequencing. Both of these methods predicted large numbers of SNPs, and the validation rate ranged from 79% to 92.5% [23]. The Universal Network-Enabled Analysis Kit (UNEAK) approach was developed for SNP discovery in switchgrass, which is a highly heterozygous polyploid (tetraploid and octoploid) species lacking a reference genome, and a total of 1.2 million putative SNPs were discovered in a diverse collection of primarily upland, northern-adapted populations [24]. In a study on hexaploid cultivated oat plants, contigs were filtered through a bioinformatics pipeline to eliminate ambiguous polymorphisms caused by subgenome homology. This procedure identified 9,448 candidate SNP loci. The greatest attrition of these candidate SNPs was based on SNP conservation between reads from a single germplasm, and 55% *in silico* SNPs were rejected [12].

Genetic linkage maps based on molecular markers can form the basis for QTL mapping and marker assistant selection and permit the elucidation of genome structure and organization. For instance, in the Tifrunner × GT-C20 cultivated peanut population, using the F_2 and F_5 generations, Wang et al. [7] and Qin et al. [25] constructed two genetic maps with 318 and 239 loci, respectively. Both genetic maps were compared to the reference consensus genetic map that was developed by Gautami et al. [26] for anchor and colinearity analysis [27]. Using the two maps and the combined multi-environment phenotyping data, Wang et al. [27] identified QTLs for thrips, tomato spotted wilt virus (TSWV), and leaf spot (LS). Although available linked markers of important traits are still lacking in peanut, we are hopeful about the future of marker-assisted breeding from its successful application of converting the peanut cultivar Tifguard [28] into 'high oleic Tifguard' in 28 mo [29] using the available limited resources in peanut.

In this study, we employed the ddRADseq approach to achieve mass discovery of SNP markers for cultivated peanut. A bioinformatics pipeline was applied for SNP calling in the parents and genotyping in the progeny. Using the newly developed markers and previously published SSR markers, a SNP-based genetic map was constructed. The characterization of this genetic map and the comparative analysis with a previous integrated consensus map were performed.

RESULTS

Library Construction and Sequencing Results

The ddRADseq protocol was used to construct reduced-representation libraries for the parents Zhonghua 5 and ICGV86699 and 166 of their RIL progenies. A rare-cutting restriction enzyme, SacI (GAGCTC), and a restriction enzyme with a more common recognition site, MseI (TTAA), were chosen based on previous success in reducing genome complexity [21]. The selected size of the DNA fragments for the ddRADseq library was 300 bp to 500 bp (with indices and adaptors). To enable multiplex sequencing of the libraries, we used a set of molecular identifying sequences (MIDs) ranging in length from 5 bp to 8 bp that allowed reads to be assigned to unique individuals. Each sequence contained adaptors, which included the sequencing primer, MID and complimentary sequence to the overhangs produced by the restriction enzymes,followed by locus-specific genomic DNA. Libraries from 12 different individuals tagged with 12 barcodes were pooled and sequenced on the Illumina HiSeq2000 platform.

Massively parallel Solexa sequencing of the ddRADseq library generated ~175 Gb of data containing 952,679,665 paired-end reads, with each read being ~90 bp in length. The Q20 (representing a quality score of 20, indicating a 1% chance of error and, thus, 99% confidence) ratio was 96.7%, and the guanine-cytosine (GC) content was 44.3%. Among these high-quality data, 72 million reads came from the parents (39,589,594 reads from Zhonghua 5 and 32,410,406 reads from ICGV86699), and ~ 1,833 million reads came from the libraries for the 166 F_9 progeny. In the RILs, the number of reads per

F_9individual ranged from 3,940,624 to18,828,436, with an average of 11,044,333 reads (Figure 1).

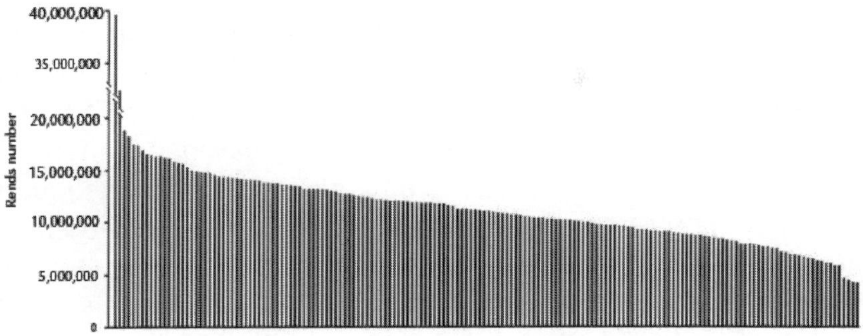

Figure 1. The numbers of sequencing reads for the RIL individuals and their parents. The first two bars of the x-axis indicate Zhonghua 5 and ICGV86699, and the following bars represent the 166 RIL lines generated from these parents; the y-axis indicates sequencing reads.

SNP Calling between the Parents

The sequencing reads of the parents were clustered using Vmatch software [30]. The number of reads forming each cluster showed eight-fold average sequencing coverage. The consensus sequences contained a total of 71,590,118 sequence tags, and the total length of the consensus sequences was 214,422,448 bp. SNP calling between the parents was performed by aligning the reads from the parents to the consensus sequences using SOAP software [31]. A total of 39,357,846 (99.4%) reads from Zhonghua 5 and 32,232,272 (99.5%) reads from ICGV86699 could be aligned to the consensus sequences. We chose uniquely mapped reads for SNP discovery. The sequences that matched more than 50 locations in the consensus sequences corresponded to 20,567 events and represented serious contaminating repetitive elements. In this case, a total of 30,977,293 (43%) reads were rejected because of multiple matching loci. Of the 40,612,825 remaining unique reads, 1,346,253 loci were eliminated because of heterozygous alleles within one parent, and 31,010 loci were removed due to less than four reads being found in each line. After applying the filtering procedure, 53,257 SNP loci between the parents were retained.

SNP Genotyping of the RIL Population

Because the construction of a SNP-based genetic map required the polymorphic markers between the two parents, the consensus sequences that did not contain SNPs were discarded, producing a reduced consensus sequences of 7,422,496 bp. Calling of SNP genotypes was performed in the population based on aligning the sequencing reads for the RIL lines to the reduced consensus sequences. A total of 516,699,812 sequencing reads from RIL individuals were aligned to the reduced consensus sequences, and the average number of aligned reads per individual was 3,112,649. Among the total aligned reads from the RIL individuals, 191,321,469 were for unique sites, and the average number of uniquely mapped reads per individual was 1,152,539, accounting for 37% of the average aligned reads for individuals. The uniquely mapped reads were chosen for subsequent SNP discovery. A total of 609,578 SNP loci were removed based on showing a heterozygous genotype, and 10,032 loci were removed due to an insufficient read depth (≤4). We detected 14,663 SNPs in the RIL population. For each individual from the RILs, the number of genotyping loci ranged from 7,606 to 10,429, averaging 8,646, and the majority of individuals presented 7750–9250 genotyping loci (Figure 2). Using a maximum missing data (MMD) threshold of 25% in the RIL population for each locus, a total of 1,765 SNP loci were finally recovered. The SNP-flanking sequences and the polymorphic sites are listed in Additional file 1.

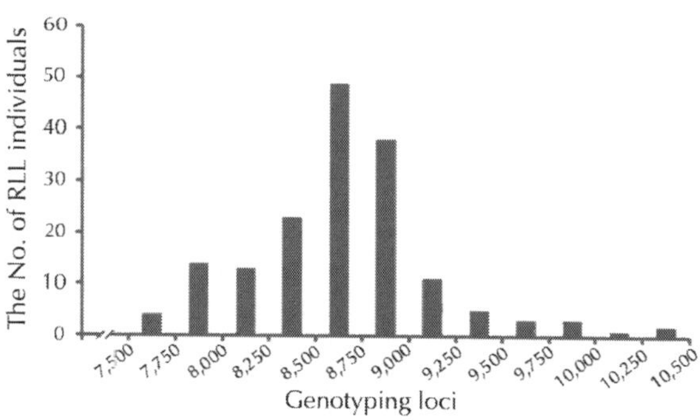

Figure 2. Distribution of genotyping loci for RIL individuals. The x-axis indicates the number of genotyping loci; the y-axis indicates the number of RIL individuals.

SNP Analysis and Validation

In total, the stringent *in silico* SNP selection criteria produced 1,765 SNPs, and the SNP distribution and the percentages of different SNP types were investigated. The SNPs were distributed evenly across the reads, with the end showing a slightly broadening range, mainly due to the decline of the base quality (Figure 3). Most of the SNPs were transition-type SNPs, with the C/T and G/A types accounting for 37% and 36% of the SNPs, respectively. The other four SNP types were transversions, which included C/G, G/T, C/A, and A/T, showing percentages ranging from 3% to 11%, accounting for 27% of all SNPs (Table 1).

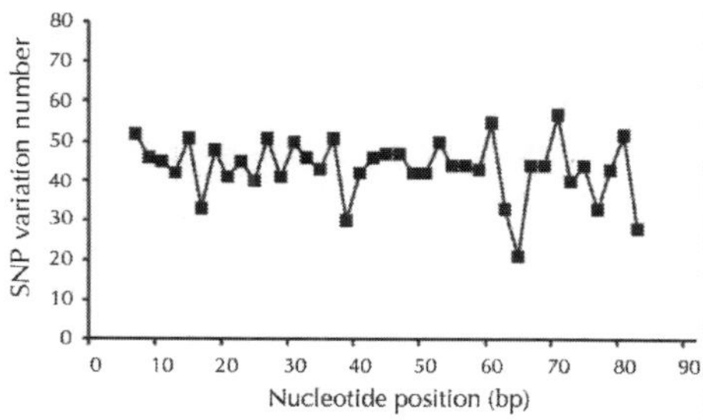

Figure 3. The distribution of the total number of SNP variations at each nucleotide position for each read.

Table 1. Statistics for the identified SNP marker types

Type of variation	Number	Proportion of type
C/G	55	3%
G/T	144	8%
C/A	186	11%
A/T	91	5%
C/T	652	37%
G/A	637	36%
Total	1765	100%

To investigate the authenticity of the identified SNPs, we randomly selected 50 SNPs for validation of single nucleotide variations. PCR primers were designed to amplify the fragments containing the SNPs.

We further sequenced the PCR products for all 50 loci amplified from the two parents using the Sanger sequencing method to confirm these SNPs. Of these 50 SNPs, 47 (94%) could be confirmed by Sanger sequencing. All 47 confirmed SNPs showed the expected nucleotide variations, while among the remaining 3 SNPs, 1 failed to amplify clearly by PCR, and 2 were a mixture of the expected allelic variations and homoeologous sequences. These results further demonstrated the efficacy of this approach for discriminating allelic SNPs from cultivated peanut.

A. hypogaea Genetic Map

Of the 1,765 developed SNP markers, 1,621 were included in the *A. hypogaea* map, which were combined into 20 linkage groups (Figure 4). To anchor and align the current map with previously published maps for peanut, 379 previously published SSR markers for single loci distributed among the 20 linkage groups of the integrated consensus map, which came from Shirasawa et al. [9] or Gautami et al. (2012) [26], were screened on the parental genotypes. As a result, 103 polymorphic markers were identified. A total of 64 markers were mapped to the 20 LGs of the current map (Table 2; Additional file 2).

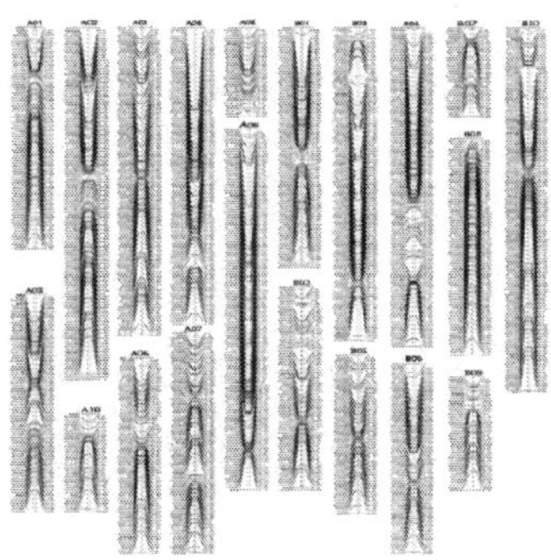

Figure 4. The SNP-based genetic linkage map for cultivated peanut using the Zhonghua 5 × ICGV86699 population. SNP markers are preceded by 'Ah-snp'. Markers are shown on right side of the LGs, while map distances are shown on the left side. Sixty-four previously published markers (underlined) were selected from the integrated consensus map of Shirasawa et al. (2013) or Gautami et al. (2012) to assign the linkage map to the corresponding chromosome.

Table 2. Characteristics of the molecular markers used for mapping

Molecular markers	Number of polymorphism primers	Number of linked markers	Number of unlinked markers	Frequency of unlinked markers (%)
SNP	1765	1621	144	8.2%
SSR	103	64	39	37.9%
Total	1926	1685	241	23.1%

Overall, the linkage map contained 1685 loci, and covered a total of 1446.7 cM, forming 1267 bins. The LGs ranged from 31.5 to 121.2 cM in length, and seven linkage groups contained over 100 marker loci. B07 and A08 were the smallest LGs, spanning 63.5 cM and 87.8 cM, respectively, and comprising 34 loci, whereas A09 was the largest LG, spanning 121.2 cM and containing 132 loci. The marker densities ranged from 0.4 cM/locus in B01 to 2.7 cM/locus in A08, resulting in an average distance of 0.9 cM between markers for the entire map (Table 3).

Table 3. Features of the 20 linkage groups

LGs	Length (cM)	No. of loci	No. of bins	Density (cM/locus)	No. of Distorted loci	No. of SDRs[a]	No. of the longest SDRs	Frequency of segregation distortion marker	Largest gap (cM)	Gaps ≤ 5
A01	76.8	83 (5)	55	0.9	25	2	12	30.1%	6.6	93.9%
A02	60.4	129 (3)	102	0.5	5	1	3	3.9%	6.5	98.4%
A03	67.9	113 (2)	73	0.6	44	4	16	38.9%	5.0	99.1%
A04	62.0	109 (2)	72	0.6	24	1	22	22.0%	17.1	98.1%
A05	66.6	80 (4)	62	0.8	43	3	25	53.8%	8.5	94.9%
A06	57.8	72 (2)	53	0.8	5	1	5	6.9%	6.3	95.8%
A07	93.0	81 (3)	71	1.2	16	3	4	19.8%	8.8	93.8%
A08	87.8	34 (3)	31	2.7	6	0	0	17.7%	9.9	84.8%
A09	121.2	132 (4)	111	0.9	14	2	4	10.6%	13.6	93.9%
A10	31.8	37 (2)	31	0.9	14	2	3	37.8%	6.2	97.2%
B01	31.5	89 (2)	67	0.4	65	8	9	73.0%	6.1	98.9%
B02	96.8	75 (3)	59	1.3	45	2	40	60.0%	10.9	90.5%
B03	74.3	115 (5)	56	0.7	28	1	22	24.4%	8.9	98.2%
B04	88.5	117 (3)	97	0.8	64	3	52	54.7%	7.7	94.8%
B05	48.5	59 (2)	47	0.8	32	4	15	54.2%	4.7	100%
B06	98.9	70 (4)	55	1.4	9	1	5	12.9%	12.9	89.9%
B07	63.5	34 (2)	29	1.9	21	2	9	61.8%	11.2	81.8%
B08	78.7	80 (4)	61	1.0	50	2	47	62.5%	6.9	94.9%
B09	48.8	43 (2)	38	1.2	36	2	26	83.7%	8.1	95.2%
B10	91.9	133 (7)	97	0.7	113	3	58	85.0%	11.6	96.2%
Total	1446.7	1685 (64)	1267	0.9	659	47	/	/	/	/
Mean	72.3	84	63	/	/	/	/	/	/	94.5%

The number in parentheses is represents the number of SSR anchor loci in the linkage group.

aSDRs, segregation distortion regions.

'Gap≤5' indicates the percentages of gaps in which the distance between adjacent markers was smaller than 5 cM.

In the map, 659 (39.1%) markers showed a skewed segregation pattern ($P<0.05$; Table 3). The segregation distortion markers were distributed among every LG. The average number and proportion of distorted markers of the LGs in the A sub-genome were 196 and 22.5%, respectively, which were lower than in the B sub-genome (463 and 56.8%, respectively; Table 3), suggesting that the chromosomal selection in the A sub-genome has smaller scale than that in the B sub-genome. The majority of the distorted markers were distributed as clusters, and 47 segregation distortion regions (SDRs) were detected and distributed in all of the linkage groups except A08. B01 had the largest number of SDRs, and B10 contained the longest SDR, which included 58 markers, spanning a distance of 24.3 cM. The degree of linkage between markers was reflected by the fact that 'Gap≤5' was observed with an average value of 94.5%. A total of 7 regions of the linkage groups contained gaps of more than 10 cM, and the largest gap in this map was 17.1 cM, located in A04 (Table 3, Figure 5).

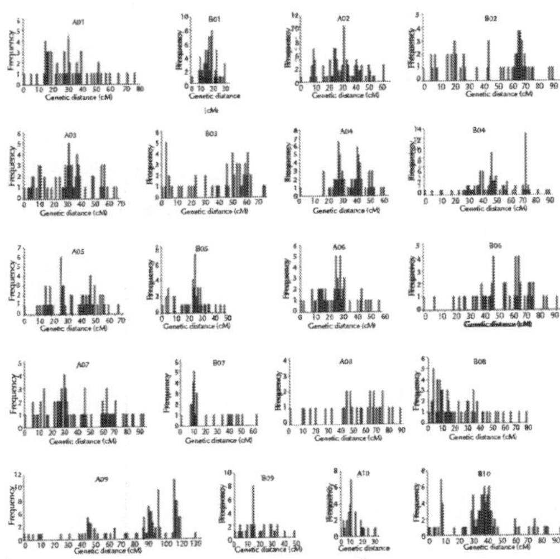

Figure 5. The X axis indicates the position in each linkage group in 1 cM intervals, and the Y axis indicates the number of bins within 1 cM.

Comparative Analysis

The linkage map in this study was aligned to the integrated consensus map developed by Shirasawa et al. [9]. The main marker types in this integrated consensus map were SSRs and transposons. In 64 single-locus SSR loci of the SNP-based linkage map, 56 were identified as having corresponding loci in the 20 chromosomes of the integrated consensus map (Additional file 3), while the remaining SSR markers were from another integrated map developed by Gautami et al. [26]. The aligned single-locus SSR loci of this SNP-based map could be treated as anchors to assign linkage groups (LGs) to specific chromosomes. Although a direct alignment of SNPs with SSRs or transposon markers is not practical, an indirect alignment of the different marker types through the GSSs sequences of peanut from NCBI is feasible. The different types of markers that map to the same sequence fragments were considered as having similar or adjacent map positions. From the alignment, 90 loci distributed on 20 linkage groups of the newly developed linkage map were identified as having corresponding loci in the integrated consensus map (Additional file 3). The corresponding LGs were collinear, except LG B03. Within the conserved regions, the orders of some of the conserved loci were altered by simple inversions or translocations. For collinear LGs, such as LG A03, two SSR markers and seven SNP markers could be mapped in the integrated consensus map, giving conserved consistent points of corresponding LG. For LG B03, 7 corresponding markers were clustered into two chromosomal segments. The first of these was roughly collinear, with 4 aligned markers spanning 18.4 cM (24.8%) on the SNP-based map and 35.5 cM (24.5%) on the integrated consensus map. The other fragment had a reversed order with 3 aligned markers spanning 12.5 cM (16.8%) on this map and 23.2 cM (16.0%) on the integrated consensus map. This observation was similar as the comparative analysis between the integrated consensus map and the TF6 population [9].

DISCUSSION

A. hypogaea is a recently formed tetraploid that most likely originated from natural hybridization of the mesopolyploids *A. duranensis* and *A. ipaensis*, which contributed to the constituent A and B genomes, respectively. In recent years, many studies of SNP development in polyploid crops have been reported. Trick et al. (2009) [32] exploited a methodology including computational tools and detected 36,424 (87.5%) hemi-SNPs and 5,169 (12.4%) simple SNPs between two rapeseed cultivars under a requirement for a minimum of four reads depth using Solexa transcriptome sequencing. Based on this study, Hu et al. (2012a, 2012b) [33, 34] developed a new method for identifying SNP markers in*Brassica napus* with filtering criteria based on the incorporation of read redundancy, quality index and lines information. Among these criteria, choosing only the unique sequences that match exactly one position in the reference genomes for SNP discovery is a particularly important filtering criterion and can greatly decrease the disturbance of paralogs within two lines. Hu et al. 2012a [33] identified 60,396 'simple SNPs', and two associated SNPs were finally mapped to a major QTL region. Hu et al. 2012b [34] detected 655 SNPs, and the validation rate reached 84%. In the present study, to decrease complexity and improve the accuracy of genotyping, we developed markers using read mapping uniqueness as a filtering criterion, too. Combined with other filtering criteria, the SNP sites were considered to be simple SNPs if there were no less than four reads depth for each genotype that revealed the same base change. In total, 53,720 SNPs were identified between the two parents, and 1,765 polymorphic markers were identified for genetic linkage map construction. Forty-seven out of 50 SNPs (94%) were verified according to Sanger sequencing, showing that the applied bioinformatics analyses were stringent and effective.

In the current study, a linkage map was finally constructed that was comprised of 1,685 marker loci, including 1,621 SNPs and 64 SSR loci, and spanning 1446.7 cM. The map was divided into 20 linkage groups and assigned to corresponding chromosomes. The first linkage map anchored to the A and B genomes was published by Foncéka et al. [5] and included 298 loci in 21 linkage groups (LGs) from a cultivated BC_1F_1 population. Because of the low marker density in the existing population-specific linkage maps and the difficulty of understanding the genomic structure of Arachis, two significant integrated consensus maps were recently constructed based on the segregation genotypes of multiple populations, anchored to 20 consensus LGs corresponding to the A and B genomes (A01-A10, B01-B10) [9, 26]. In this study, the applied SSR markers amplified single loci, distributed among the 20 linkage groups from the above two integrated consensus maps. The subsequent linkage analysis generated a total of 20 linkage groups. The present linkage map corresponds to the number of chromosomes (n = 20) in cultivated peanut, and the linkage groups were assigned to the specific chromosomes.

Segregation distortion is a common biological phenomenon and is one of the engines driving evolutionary processes. It can be observed in almost all types of hybrid segregating populations. In general, the skewed segregation ratio of RIL populations is higher than that of backcross populations (BC) and doubled haploid populations (DH). F_2 populations show the lowest skewed segregation ratio [35]. The genetic basis of segregation distortion is still under debate, and gametophyte and/or zygotic selection and chromosomal rearrangements may be the main cause of this phenomenon. Studies have demonstrated that a large number of segregation distortions and SDRs occur in many species, such as maize [36], barley [37], and potato [38]. In this study, we used a RIL F_9 population as a mapping population to construct a linkage map, and 659 (39.1%) markers showed skewed segregation. This high-generation population could improve the accuracy of bioinformatics analysis for SNP discovery because of long stretches of consecutive homozygous genotypes, while the marker more likely skewed segregation probably related to the many generations of natural selection and artificial sampling involved in the construction of the RIL population. In this map, most of the markers exhibiting segregation distortion were distributed as clusters in linkage groups. Distorted markers were often strung together, suggesting that there has been selection for gametophytes or sporophytes.

As discussed above, both the SNP and SSR markers on this map presented single-locus nature in the AABB genome. Comparative analysis between the AA and between BB genomes were performed and showed that all LGs in the SNP-based map were collinear with their corresponding LGs in the integrated consensus map, except LG B03, for which the corresponding markers were clustered into two chromosomal segments and had reversed orders. This observation suggested the chromosome segment with inversion or rearrangement in LG B03. Relative to the large peanut genome, the number of markers is still low and the available peanut sequence is limited and the common GSSs that can be used as bridges to align SNP and SSR markers less. The completion of peanut genome sequence and the development of increasing numbers of molecular markers will establish more alignment points between the genetic maps with different types of markers. Even so, the alignment of some parts of the present map with integrated consensus maps of peanut demonstrates the possibility of developing SNP markers for constructing linkage groups in cultivated peanut and improving our understanding of the genome.

The current version of the cultivated peanut linkage map is a considerable improvement compared with the previously available versions (Table 4). There are two major reasons for this improvement. First, this is the only SNP-based linkage map that has been produced for cultivated peanut. Initially, genetic maps were developed for wild species with AA- and BB- genomes. For cultivated peanut species or crosses of cultivated and synthetic tetraploid peanut species, a few linkage maps have recently been constructed (Table 4), and some of these maps were based on multiple populations. Earlier maps used RFLP or AFLP markers, while the later maps were mainly based on SSR markers. Varshney et al. [39] constructed the first SSR linkage map for cultivated peanut. Since that time, the construction of SSR-based genetic linkage maps for *A. hypogaea* has proceeded rapidly. This study was the first to develop SNP markers on a large scale to construct a genetic map for cultivated peanut. Another obvious improvement is that the maximum number of markers for a linkage map in a single mapping population was used. Shirasawa et al. [8] published a high-density genetic map composed of 1,114 loci, including SSR and transposon markers. Another high-density map included 1,469 loci, with an average distance of 1.0 cM between adjacent loci [9]. The map produced in the present study contains 1,685 markers, and the average genetic interval is 0.9 cM per marker. To our knowledge, the number of markers included in this map is the highest among the available population-specific linkage maps for tetraploid peanuts (Table 4).

Table 4. Comparison of tetraploid linkage maps for Arachis from a single population

Cross combination	Population type	Types of markers	Number of markers	Maps length(cM)	Groups	References
A. hypogaea 'Florunner' × (*A. batizocoi* 'K9484' × (*A. cardenasii*'GKP10017' × *A. diogoi*' GKP10602') $^{4\times}$)	BC_1F_1	RFLP	370	2210	23	[4]
A. hypogaea 'ICG12991' × *A. hypogaea* 'ICGV-SM 93541'	F_2	AFLP	12	139.4	5	[40]
A. hypogaea 'TAG24' × *A. hypogaea* 'ICGV 86031'	RIL	SSR	135	1270.5	22	[39]
A. hypogaea 'Fleur11' × (*A. ipaënsis* 'KG30076' × *A. duranensis*'V14167')$^{4\times}$	BC_1F_1	SSR	298	1843.7	21	[5]
A. hypogaea 'Yueyou 13' × *A. hypogaea* 'Zhen Zhuhei'	RIL	SSR	132	684.9	19	[41]
A. hypogaea 'Yueyou 13' × *A. hypogaea* 'FU 95-5'	RIL	SSR	109	540.69	21	[41]
A. hypogaea 'Yueyou 13' × *A. hypogaea* 'J 11'	RIL	SSR	46	401.7	13	[41]
A. hypogaea 'TAG 24' × *A. hypogaea* 'GPBD 4'	RIL	SSR	188	1,922.4	20	[6]
A. hypogaea 'TG 26' × *A. hypogaea* 'GPBD 4'	RIL	SSR	181	1,963	21	[6]
A. hypogaea 'ICGS76' × *A. hypogaea* 'CSMG 84-1'	RIL	SSR	119	2,208.2	20	[42]
A. hypogaea 'ICGS 44' × *A. hypogaea* 'ICGS 76'	RIL	SSR	82	831.4	15	[42]
A. hypogaea 'SunOleic 97R' × *A. hypogaea* 'NC94022'	RIL	SSR, CAPs	172	920.7	22	[25]
A. hypogaea 'Tifrunner' × *A. hypogaea* 'GT-C20'	F_2	SSR	318	1674.4	21	[7]
A. hypogaea 'YI-0311' × *A. hypogaea* 'Nakateyutaka'	F_2	SSR, transposon, SNP	326	1332.9	19	[8]
A. hypogaea 'Satonoka' × *A. hypogaea* 'Kintoki'	F_2	SSR, transposon	1114	2166.4	21	[8]
A. hypogaea 'Runner IAC 886' × c*A. ipaënsis* 'K30076' × *A. duranensis* 'V14167')$^{4\times}$	RIL	SSR, transposon	1469	1442	20	[9]
A. hypogaea 'Zhonghua 5' × *A. hypogaea* 'ICGV86699'	RIL	SNP, SSR	1685	1441.1	20	This study

Molecular markers and genetic linkage maps are the pre-requisites for undertaking genetic mapping of important traits and molecular breeding activities in crops. The female parent of the RIL population Zhonghua 5 is a popular high-yield cultivar in China, but it is susceptible to late leaf spot. However, the male parent, ICGV86699, has excellent resistance to this disease (Additional file 4), which is the most widely distributed peanut disease in China. The tools generated in this study will accelerate the genetic research and the process of introgression of beneficial traits into preferred varieties of cultivated peanut, such as resistance to late leaf spot. Because the high-density linkage groups were constructed based on molecular markers developed at the whole-genome level, it will also serve as a reference for positioning sequence scaffolds on the physical map to assist in the assembly of the peanut genome sequence.

CONCLUSIONS

In this study, we constructed RRLs for two parents and 166 of their RIL progenies using the ddRADseq technique. Combined with a next-generation sequencing approach, we detected SNPs in cultivated peanut through the adoption of appropriate filtering criteria and constructed a genetic map containing 1,621 SNP loci and 64 SSR loci distributed among 20 LGs. All LGs in the SNP-based map were collinear with their corresponding LGs in the integrated map, except B03, where chromosome segment inversions or rearrangements maybe involved. The results of this study will provide a useful resource for molecular markers, QTL mapping, molecular breeding, and facilitating the assembly of a reference genome sequence for the peanut.

METHODS

Plant Material

A RIL population including 166 F_9 lines was developed from a cross between Zhonghua 5 and ICGV86699. The parent Zhonghua 5 is an early maturing, high-yield popular cultivar but susceptible to late leaf spot disease. The parent ICGV86699 is a breeding variety from strains of distant hybridization, and it has resistance to late leaf spot that was introgressed to *A. hypogaea* from *A. duranensis*. The population was developed in the experimental field of the Oil Crops Research Institute of the Chinese Academy of Agricultural Sciences in Wuhan, Hubei Province. Genomic DNA was extracted from young leaf tissue essentially as described by Grattapaglia and Sederoff (1994) [43].

ddRADseq Library Construction and Sequencing

The procedure was performed as described by Chen et al. (2013) [21] with some modifications. First, genomic DNA was double digested separately with restriction enzymes. The double digest reactions were carried out in a volume of 25 μl containing approximately 150 ng of genomic DNA, 5 U of SacI and MseI (Fermentas), and 1× buffer. The reaction mixture was incubated at 37°C for 6 hr and 65°C for 90 min. Second, the fragments were ligated with adaptors. The ligation reaction was conducted in a reaction volume of 50 μl at 16°C overnight, containing 10 pmol of SacI and MseI adaptors, and 1,000 U of T4 DNA Ligase (New England Biolabs [NEB]). To ensure that the digestion was complete, the digestions were performed again with the same enzymes. Each sample was then amplified via PCR in a 50 μl reaction volume, containing 50–100 ng of adaptor-ligated DNA fragments as a template, 1× HF buffer, 3.5 mM $MgCl_2$, 0.4 mM dNTPs, 0.5 U of iProof polymerase (Bio-Rad), and 4 pmol of two overhang primers. PCR amplification was performed according to the following program: 98°C for 2 min, followed by 13 cycles at 98°C for 30 s, 60°C for 30 s, and 72°C for 15 s, and a final extension at 72°C for 5 min. The PCR products were run on a 2% agarose gel, and fragments of 300–500 bp were recovered from the gel. The samples from 12 individuals were pooled together, and DNA was isolated using a Gel Extraction Kit (Qiagen). The libraries were quantified using Qubit fluorometer (Invitrogen), Agilent 2100 (Agilent Technologies) and real-time quantitative PCR, then submitted for sequencing on the Illumina HiSeq2000 platform.

In silico SNP Identification and Genotyping

The bioinformatics process used for the identification of SNP markers is presented in Figure 6. Based on the Illumina raw data, a custom Perl script was written to sort sequences from individual samples based on indexes and trimmed barcode sequences for faster processing. Only sequences that presented an exact match to a barcode, followed by the expected sequence of nucleotides remaining after a SacI or MseI cut site were retained. The low-quality, contaminant sequences were trimmed using NGS QC Toolkit [44].

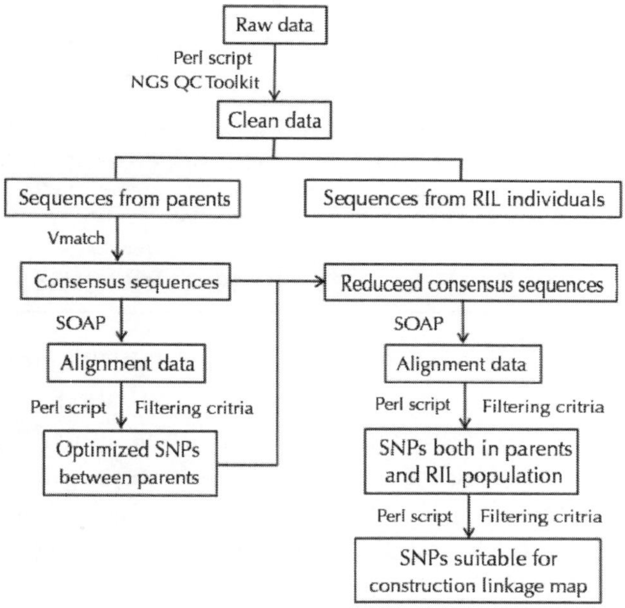

Figure 6. Bioinformatics pipeline for SNP discovery.

The cleaned data were clustered with Vmatch at a stringent level, where the default parameter setting was used, as applied in a number of SNP mining programs [30, 45]. Calling of single nucleotide polymorphisms (SNPs) was based on the alignment of the parental sequences to the consensus sequences using SOAP software [31]. Then, Custom Perl scripts were used for SNP calling according to published reports [46, 47]. The SNP calling fulfilled the following criteria: 1) to exclude regions of complex polymorphism, all PE reads from each line were aligned to the consensus sequences with at most two nucleotide mismatches on each strand of a read; 2) to avoid paralogue interference, only uniquely aligned reads were selected; 3) to avoid non-uniform polymorphisms, nucleotide variations present a 100% frequency within a genotype; and 4) to assure accuracy, every allele had to present a sequencing depth of no less than four reads. After identifying SNPs between parents, the SNP-containing sequences were extracted from the consensus sequences, thus producing reduced consensus sequences. For SNP detection in the RIL population, the same filtering criteria were used as in the parents. We calculated the likelihood of each line's genotype using SOAPsnp [31]. A Bayesian model was applied, and the genotype with the highest probability was selected as the genotype of the individual at the specific locus. Each marker was required to have an allele present in at least 75% of F_9 individuals, and each allele had to be present in at least 30 F_9 individuals. Marker genotypes not meeting the minimum thresholds were scored as missing data.

SNP Validation through Resequencing

Primer3plus was used to design primers to amplify the target fragments including the SNP variations. The SNPs that were validated between the two parents were subjected to genotype analysis in the RIL population. PCR amplifications were carried out in a volume of 20 μl, containing 100 ng of DNA template, 1 × Pfu buffer, 4 mM $MgCl_2$, 0.4 mM dNTPs, 5 pmol of each primer, and 0.4 U of Pfu. Thermocycling was performed at 94°C for 3 min, followed by 35 cycles of 94°C for 30 s, 60°C for 1 min and 72°C for 45 s, with a final extension step of 72°C for 5 min, and then holding at 4°C. Aliquots (5 μl) of the PCR products were first analyzed on agarose gels to verify successful amplification, and the remaining PCR products were directly sequenced by BGI using an ABI3730 sequencer.

Genetic Linkage Map Construction

The RIL F_9 population, consisting of 166 individuals, was utilized to construct a genetic map. The SNP marker sequences that were used for the genetic map are listed in Additional file 1: Table S1. The input datasets were constructed from 1,765 genotyped SNP markers and 103 previously published SSR loci. The program Joinmap 4.0 [48] was used to calculate the marker order and genetic distance. Recombination frequencies ≤ 0.45 and LOD scores ≥ 2.0 were used to create groups.

The Kosambi mapping function was employed for map length estimations. Markers were tested for segregation distortion by the chi-square test. A graphic representation of the map was generated using Mapchart 2.0 software [49].

ACKNOWLEDGEMENTS

This research was supported by the Major State Basic Research Development Program of China (973 Program) (grant no. 2011CB109304), the National Natural Science Foundation of China (grants no. 31301362 and 31271764), the National Program for Crop Germplasm Protection of China (grant no. 2005DKA21002-13).

REFERENCES

1. Bennett MD, Bhandol P, Leitch IJ. Nuclear DNA amounts in angirosperms and their modern uses-807 new estimates. Ann Bot. 2000;86:859–909. doi: 10.1006/anbo.2000.1253.
2. Stalker HT, Mozingo LG. Molecular markers of Arachis and marker-assisted selection. Peanut Sci.2001;28:117–123. doi: 10.3146/i0095-3679-28-2-13.
3. Paterson AH, Stalker HT, Gallo-Meagher M, Burow MD, Dwivedi SL, Crouch JH, Mace ES. Genomics and genetic enhancement of peanut. In: Wilson RF, Stalker HT, Brummer CE, editors.Genomics for Legume Crops. Champaign IL: Amer Oil Chem Soc; 2004. pp. 97–109.
4. Burow MD, Simpson CE, Starr JL, Paterson AH. Transmission genetics of chromatin from a synthetic amphiploid in cultivated peanut (A. hypogaea L.): broadening the gene pool of a monophyletic polyploid species. Genetics. 2001;159:823–37.
5. Foncéka D, Hodo-Abalo T, Rivallan R, Faye I, Sall MN, Ndoye O, Fávero AP, Bertioli DJ, Glaszmann J-C, Courtois B, Rami J-F. Genetic mapping of wild introgressions into cultivated peanut: a way toward enlarging the genetic basis of a recent allotetraploid. BMC Plant Biol. 2009;9:103. doi: 10.1186/1471-2229-9-103.

6. Sujay V, Gowda MV, Pandey MK, Bhat RS, Khedikar YP, Nadaf HL, Gautami B, Sarvamangala C, Lingaraju S, Radhakrishan T, Knapp SJ, Varshney RK. Quantitative trait locus analysis and construction of consensus genetic map for foliar disease resistance based on two recombinant inbred line populations in cultivated groundnut (Arachis hypogaea L.) Mol Breed. 2012;30:773–88. doi: 10.1007/s11032-011-9661-z.
7. Wang H, Penmetsa RV, Yuan M, Gong L, Zhao Y, Guo B, Farmer AD, Rosen BD, Gao J, Isobe S, Bertioli DJ, Varshney RK, Cook DR, He G. Development and characterization of BAC-end sequence derived SSRs, and their incorporation into a new higher density genetic map for cultivated peanut (Arachis hypogaea L.) BMC Plant Biol. 2012;12:10. doi: 10.1186/1471-2229-12-10.
8. Shirasawa K, Koilkonda P, Aoki K, Hirakawa H, Tabata S, Watanabe M, Hasegawa M, Kiyoshima H, Suzuki S, Kuwata C, Naito Y, Kuboyama T, Nakaya A, Sasamoto S, Watanabe A, Kato M, Kawashima K, Kishida Y, Kohara M, Kurabayashi A, Takahashi C, Tsuruoka H, Wada T, Isobe S. In silico polymorphism analysis for the development of simple sequence repeat and transposon markers and construction of linkage map in cultivated peanut. BMC Plant Biol. 2012;12:80. doi: 10.1186/1471-2229-12-80.
9. Shirasawa K, Bertioli DJ, Varshney RK, Moretzsohn MC, Leal-Bertioli SCM, Thudi M, Pandey MK, Rami J-F, Foncéka D, Gowda MVC, Qin H, Guo B, Hong Y, Liang X, Hirakawa H, Tabata S, Isobe S. Integrated consensus map of cultivated peanut and wild relatives reveals structures of the A and B genomes of Arachis and divergence of the legume genomes. DNA Res. 2013;20(2):173–84. doi: 10.1093/dnares/dss042.
10. Nagy ED, Guo Y, Tang S, Bowers JE, Okashah RA, Taylor CA, Zhang D, Khanal S, Heesacker AF, Khalilian N, Farmer AD, Carrasquilla-Garcia N, Penmetsa RV, Cook D, Stalker HT, Nielsen N, Ozias-Akins P, Knapp SJ. A high-density genetic map of Arachis duranensis, a diploid ancestor of cultivated peanut. BMC Genomics. 2012;13:469. doi: 10.1186/1471-2164-13-469.
11. Brooks AJ. The essence of SNPs. Gene. 1999;234:177–186. doi: 10.1016/S0378-1119(99)00219-X.

12. Oliver RE, Lazo GR, Lutz JD, Rubenfield MJ, Tinker NA, Anderson JM, Morehead NHW, Adhikary D, Jellen EN, Maughan PJ, Guedira GLB, Chao S, Beattie AD, Carson ML, Rines HW, Obert DE, Bonman JM, Jackson EW. Model SNP development for complex genomes based on hexaploid oat using high-throughput 454 sequencing technology. BMC Genomics. 2011;12:77. doi: 10.1186/1471-2164-12-77.
13. Craig DW, Pearson JV, Szelinger S, Sekar A, Redman M, Corneveaux JJ, Pawlowski TL, Laub T, Nunn G, Stephan DA, Homer N, Huentelman MJ. Identification of genetic variants using barcoded multiplexed sequencing. Nat Methords. 2008;5(10):887–893. doi: 10.1038/nmeth.1251.
14. Huang X, Feng Q, Qian Q, Zhao Q, Wang L, Wang A, Guan J, Fan D, Weng Q, Huang T, Dong G, Sang T, Han B. High-throughput genotyping by whole-genome resequencing. Genome Res.2009;19(6):1068–1076. doi: 10.1101/gr.089516.108.
15. Xie W, Feng Q, Yu H, Huang X, Zhao Q, Xing Y, Yu S, Han B, Zhang Q. Parent-independent genotyping for constructing an ultrahigh-density linkage map based on population sequencing. Proc Nat Acad Sci USA. 2010;107(23):10578–10583. doi: 10.1073/pnas.1005931107.
16. Pfender WF, Saha MC, Johnson EA, Slabaugh MB. Mapping with RAD (restriction-site associated DNA) markers to rapidly identify QTL for stem rust resistance in Lolium perenne. Theor Appl Genet.2011;122:1467–1480. doi: 10.1007/s00122-011-1546-3.
17. Huang X, Zhao Y, Wei X, Li C, Wang A, Zhao Q, Li W, Guo Y, Deng L, Zhu C, Fan D, Lu Y, Weng Q, Liu K, Zhou T, Jing Y, Si L, Dong G, Huang T, Lu T, Feng Q, Qian Q, Li J, Han B. Genome-wide association study of flowering time and grain yield traits in a worldwide collection of rice germplasm. Nat Genet. 2012;44(1):32–39. doi: 10.1038/ng.1018.
18. Baird NA, Etter PD, Atwood TS, Currey MC, Shiver AL, Lewis ZA, Selker EU, Cresko WA, Johnson EA. Rapid SNP discovery and genetic mapping using sequenced RAD markers. PLoS One.2008;3(10):e3376. doi: 10.1371/journal.pone.0003376.
19. Chutimanitsakun Y, Nipper RW, Cuesta-Marcos A, Cistue L, Corey A, Filichkina T, Johnson EA, Hayes PM. Construction and application for QTL analysis of a restriction site associated DNA

(RAD) linkage map in barley. BMC Genomics. 2011;12:4. doi: 10.1186/1471-2164-12-4.

20. Peterson BK, Weber JN, Kay EH, Fisher HS, Hoekstra HE. Double digest RADseq: an inexpensive method for De Novo SNP discovery and genotyping in model and non-model species. PLoS One.2012;7(5):e37135. doi: 10.1371/journal.pone.0037135.
21. Chen X, Li X, Zhang B, Xu J, Wu Z, Wang B, Li H, Younas M, Huang L, Luo Y, Wu J, Hu S, Liu K. Detection and genotyping of restriction fragment associated polymorphisms in polyploid crops with a pseudo-reference sequence: a case study in allotetraploid Brassica napus. BMC Genomics.2013;14:346. doi: 10.1186/1471-2164-14-346.
22. Recknagel H, Elmer KR, Meyer A. A hybrid genetic linkage map of two ecologically and morphologically divergent midas cichlid fishes (Amphilophus spp.) obtained by massively parallel DNA sequencing (ddRADSeq) Genetics. 2013;3:65–74.
23. Hyten DL, Cannon SB, Song Q, Weeks N, Fickus EW, Shoemaker RC, Specht JE, Farmer AD, May GD, Cregan PB. High-throughput SNP discovery through deep resequencing of a reduced representation library to anchor and orient scaffolds in the soybean whole genome sequence. BMC Genomics. 2010;11:38. doi: 10.1186/1471-2164-11-38.
24. Lu F, Lipka AE, Glaubitz J, Elshire R, Cherney JH, Casler MD, Buckler ES, Costich DE. Switchgrass Genomic Diversity, Ploidy, and Evolution: novel Insights from a Network-Based SNP Discovery Protocol. PLoS Genet. 2013;9(1):e1003215. doi: 10.1371/journal.pgen.1003215.
25. Qin H, Feng S, Chen C, Guo Y, Knapp S, Culbreath A, He G, Wang ML, Zhang X, Holbrook CC, Ozias-Akins P, Guo B. An integrated genetic linkage map of cultivated peanut (*Arachis hypogaea* L.) constructed from two RIL populations. Theor Appl Genet. 2012;124:653–64. doi: 10.1007/s00122-011-1737-y.
26. Gautami B, Fonce´ka D, Pandey MK, Moretzsohn MC, Sujay V, Qin H, Hong Y, Faye I, Chen X, Prakash AB, Shah TM, Gowda MVC, Nigam SN, Liang X, Hoisington DA, Guo B, Bertioli DJ, Rami J-F, Varshney RK. An international reference consensus genetic map with 897 marker loci based on 11 mapping

populations for tetraploid Groundnut (Arachis hypogaea L.) PLoS ONE. 2012;7:e41213. doi: 10.1371/journal.pone.0041213.

27. Wang H, Manish KP, Qiao L, Qin H, Culbreath AK, He G, Varshney RK, Scully BT, Guo B. Genetic Mapping and Quantitative Trait Loci Analysis for Disease Resistance Using F_2 and F_5Generation-based Genetic Maps Derived from 'Tifrunner' × 'GT-C20' in Peanut. The Plant Genome.2013;6:3. doi: 10.3835/plantgenome2013.05.0018.
28. Holbrook CC, Timper P, Culbreath AK, Kvien CK. Registration of 'Tifguard' peanut. J Plant Reg.2008;2:92–94. doi: 10.3198/jpr2007.12.0662crc.
29. Chu Y, Wu CL, Holbrook CC, Tillman BL, Person G, Ozias-Akins P. Marker-assisted selection to pyramid nematode resistance and the high oleic trait in peanut. Plant Gen. 2011;4:110–117. doi: 10.3835/plantgenome2011.01.0001.
30. Willing E-M, Hoffmann M, Klein JD, Weigel D, Dreyer C. Paired-end RAD-seq for de novo assembly and marker design without available reference. Bioinformatics. 2011;27:2187–2193. doi: 10.1093/bioinformatics/btr346.
31. Li R, Yu C, Li Y, Lam T, Yiu S, Kristiansen K, Wang J. SOAP2: an improved ultrafast tool for short read alignment. Bioinformatics. 2009;25:1966–1967. doi: 10.1093/bioinformatics/btp336.
32. Trick M, Long Y, Meng J, Bancroft I. Single nucleotide polymorphism (SNP) discovery in the polyploidy Brassica napus using Solexa transcriptome sequencing. Plant Biotechnol J. 2009;7:334–346. doi: 10.1111/j.1467-7652.2008.00396.x.
33. Hu Z, Hua W, Huang S, Yang H, Zhan G, Wang X, Liu G, Wang H. Discovery of pod shatter-resistant associated SNPs by deep sequencing of a representative library followed by bulk segregant analysis in rapeseed. PLoS ONE. 2012;7(4):e34253. doi: 10.1371/journal.pone.0034253.
34. Hu Z, Huang S, Sun M, Wang H, Hua W. Development and application of single nucleotide polymorphism markers in the polyploid Brassica napus by 454 sequencing of expressed sequence tags.Plant Breeding. 2012;131:293–299. doi: 10.1111/j.1439-0523.2011.01947.x.

35. Wang W, Huang S, Liu Y, Fang Z, Yang L, Hua W, Yuan S, Liu S, Sun J, Zhuang M, Zhang Y, Zeng A. Construction and analysis of a high-density genetic linkage map in cabbage (Brassica oleracea L. var. capitata) BMC Genomics. 2012;13:523. doi: 10.1186/1471-2164-13-523.
36. Lu H, Romero-Severson J, Bernardo R. Chromosomal regions associated with segregation distortion in maize. Theor Appl Genet. 2002;105:622–628. doi: 10.1007/s00122-002-0970-9.
37. Li H, Kilian A, Zhou M, Wenzl P, Huttner E, Mendham N, McIntyre L, Vaillancourt RE. Construction of a high-density composite map and comparative mapping of segregation distortion regions in barley. Mol Genet Genomics. 2010;284:319–331. doi: 10.1007/s00438-010-0570-3.
38. Tai GCC, Seabrook JEA, Aziz AN. Linkage analysis of anther-derived monoploids showing distorted segregation of molecular markers. Theor Appl Genet. 2000;101:126–130. doi: 10.1007/s001220051460.
39. Varshney RK, Bertioli DJ, Moretzsohn MC, Vadez V, Krishnamurthy L, Aruna R, Nigam SN, Moss BJ, Seetha K, Ravi K, He G, Knapp SJ, Hoisington DA. The first SSR-based genetic linkage map for cultivated groundnut (Arachis hypogaea L.) Theor Appl Genet. 2009;118:729–39. doi: 10.1007/s00122-008-0933-x.
40. Herselman LR, Thwaites FM, Kimmins B, Courtois PJA, Merwe VD, Seal SE. Identification and mapping of AFLP markers linked to peanut (Arachis hypogaea L.) resistance to the aphid vector of groundnut rosette disease. Theor Appl Genet. 2004;109:1426–33. doi: 10.1007/s00122-004-1756-z.
41. Hong Y, Chen X, Liang X, Liu H, Zhou G, Li S, Wen S, Holbrook CC, Guo B. A SSR-based composite genetic linkage map for the cultivated peanut (Arachis hypogaea L.) genome. BMC Plant Biol. 2010;10:17. doi: 10.1186/1471-2229-10-17.
42. Gautami B, Pandey MK, Vadez V, Nigam SN, Ratnakumar P, Krishnamurthy L, Radhakrishnan T, Gowda MVC, Narasu ML, Hoisington DA, Knapp SJ, Varshney RK. Quantitative trait locus analysis and construction of consensus genetic map for drought tolerance traits based on three recombinant inbred line populations in cultivated groundnut (Arachis hypogaea L.) Mol Breeding. 2012;30:773–88. doi: 10.1007/s11032-011-9660-0.

43. Grattapaglia D, Sederoff R. Genetic linkage maps of Eucalyptus grandis and Eucalyptus urophylla using a pseudo-testcross: mapping strategy and RAPD markers. Genetics. 1994;137:1121–37.
44. Patel RK, Jain M. NGS QC Toolkit: A Toolkit for Quality Control of Next Generation Sequencing Data. PLoS ONE. 2012;7(2):e30619. doi: 10.1371/journal.pone.0030619.
45. Chong Z, Ruan J, Wu C-I. Rainbow: an integrated tool for efficient clustering and assembling RAD-seq reads. Bioinformatics. 2012;28(21):2732–2737. doi: 10.1093/bioinformatics/bts482.
46. Kumar S, Banks TW, Cloutier S. SNP Discovery through next-generation sequencing and its applications. Int J Plant Genomics. 2012;2012:1–15.
47. Bancroft I, Morgan C, Fraser F, Higgins J, Wells R, Clissold L, Baker D, Long Y, Meng J, Wang X, Liu S, Trick M. Dissecting the genome of the polyploid crop oilseed rape by transcriptome sequencing. Nat Biotechnol. 2011;29(8):762–766. doi: 10.1038/nbt.1926.
48. Van Ooijen JW. JoinMap 4, Software for the calculation of genetic linkage maps in experimental populations. Kyazma BV, Netherlands: Wageningen; 2006.
49. Voorrips RE. MapChart: software for the graphical presentation of linkage maps and QTLs. J. Hered. 2002;93:77–8. doi: 10.1093/jhered/93.1.77.

Citations

CHAPTER 1

Yang Zhou, Jin-shui Wang, Xiao-jia Yang, et al., Peanut Allergy, Allergen Composition, and Methods of Reducing Allergenicity: A Review, International Journal of Food Science, vol. 2013, Article ID 909140, 8 pages, 2013. doi:10.1155/2013/909140

CHAPTER 2

L. Sun, R. Hu, G. Shen and H. Zhang, Genetic Engineering Peanut for Higher Drought- and Salt-Tolerance, Food and Nutrition Sciences, Vol. 4 No. 6A, 2013, pp. 1-7. doi: 10.4236/fns.2013.46A001

CHAPTER 3

C. Corley Holbrook, Peggy Ozias-Akins, Ye Chu and Baozhu Guo, Impact of Molecular Genetic Research on Peanut Cultivar Development, doi:10.3390/agronomy1010003

CHAPTER 4

Chuan Tang Wang, Yue Yi Tang, Xiu Zhen Wang, Qi Wu, Hua Yuan Gao, Tong Feng, Jun Wei Su, Shu Tao Yu, Xian Lan Fang, Wan Li Ni, Yan Sheng Jiang, Lang Qian and Dong Qing Hu (2012). Mutagenesis: A Useful Tool for the Genetic Improvement of the Cultivated Peanut (Arachis hypogaea L.), Mutagenesis, Dr. Rajnikant Mishra (Ed.), ISBN: 978-953-51-0707-1, InTech, DOI: 10.5772/50514.

CHAPTER 5

Soraya C. M. Leal-Bertioli1, Silvio P. Santos, Karinne M. Dantas, Peter W. Inglis, Stephan Nielen, Ana C. G. Araujo1, Joseane P. Silva1, Uiara Cavalcante, Patricia M. Guimarães, Ana Cristina M. Brasileiro1, Noelia Carrasquilla-Garcia, R. Varma Penmetsa, Douglas Cook, Márcio C. Moretzsohn and David J. Bertioli, Arachis Batizocoi: A Study of its Relationship to Cultivated Peanut (A. hypogaea) and its Potential for Introgression of wild genes into the Peanut crop using Induced Allotetraploids, doi: 10.1093/aob/mcu237

CHAPTER 6

Suping Feng Xingjun Wang, Xinyou Zhang, Phat M. Dang, C. Corley Holbrook, Albert K. Culbreath, Yaoting Wu, and Baozhu Guo, Peanut (Arachis hypogaea) Expressed Sequence Tag Project: Progress and Application, http://dx.doi.org/10.1155/2012/373768

CHAPTER 7

Pasupuleti Janila, S. N. Nigam1 Manish K. Pandey P. Nagesh1 and Rajeev K. Varshney, Groundnut Improvement: Use of Genetic and Genomic Tools, doi: 10.3389/fpls.2013.00023

CHAPTER 8

Kenta Shirasawa, Sachiko Isobe, Satoshi Tabata, and Hideki Hirakawa, Kazusa Marker DataBase: A Database for Genomics, Genetics, and Molecular breeding in Plants, doi:10.1270/jsbbs.64.264

CHAPTER 9

Shivali Sharma, H. D. Upadhyaya, R. K. Varshney, and C. L. L. Gowda, Pre-Breeding for Diversification of Primary gene pool and Genetic Enhancement of Grain Legumes, doi: 10.3389/fpls.2013.00309

CHAPTER 10

Xiaojing Zhou, Youlin Xia, Xiaoping Ren, Yulin Chen, Li Huang, Shunmou Huang, Boshou Liao, Yong Lei, Liyin Yan, and Huifang Jiang, Construction of a SNP-based Genetic Linkage map in Cultivated Peanut based on large scale Marker Development using next-Generation Double-Digest Restriction-site-Associated DNA Sequencing (ddRADseq), doi: 10.1186/1471-2164-15-351

Index

A

Agriculture 118
Agronomical Plants 191, 193
Allotetraploids 82, 83, 87, 88, 90, 95, 96, 97, 98, 99, 100, 101, 107, 108, 109
Amplified Fragment Length Polymorphism (Aflp) 192
Angiosperms 118

B

Biosynthesis 25, 26
Branches 90, 104, 106
Breeding 82, 83, 105, 109, 110

C

Chickpea 214, 215, 216, 218, 222, 223, 224, 226, 228, 229, 231, 233, 234, 235, 238, 240, 241, 242, 243, 244, 245, 246, 247, 248, 250, 251, 252, 253, 254
Cloning 192, 204
Crop 143, 144, 146, 147, 148, 153, 156, 157, 159, 171, 176, 186
Cytogenetic Analysis 90, 103

D

Diploid Wild 45, 51

Doubled Haploid (Dh) 258
Droughtand 24, 25, 29, 31

E

Enzymatic Treatment 2, 13, 14, 15
Eucalyptus 192, 194, 204, 205
Expressed Sequence Tag (Est) 119, 132

F

Fermentation 2, 14, 22
Food Allergens 119
Food Industry 2, 9

G

Genetic Base 213, 215, 220, 222, 230, 233, 235, 238, 239, 244, 245
Genetic Engineering 2
Genetic Resources 213, 215, 221, 241, 246, 251
Genome 256, 257, 258, 259, 260, 267, 270, 273, 279, 280, 282, 283
Genome In Situ Fluorescent Hybridization (Gish) 82, 85
Genome-Wide Association 144, 171
Global Production 118
Global Seed Vault 216
Grain Legumes 213, 214, 215, 216, 220, 233, 237, 239, 241, 248
Groundnut Vi, 143, 144, 145, 146, 147, 150, 153, 155, 158, 160, 161, 165, 168, 172, 174, 178, 179, 180, 181, 182, 184, 185, 186, 187, 189, 287
Guanine-Cytosine (Gc) 260

H

Health Benefits 119
Health Problem 2
Hybrid Plants 86
Hybrids 82, 83, 84, 85, 87, 88, 90, 95, 99, 100, 101, 102, 107, 108, 109

I

Influencing 256
International Crops Research Institute For The Semi-Arid Tropics (Icrisat) 214, 215, 244
Irradiation Breeding 2
Irrigation 24, 27, 28, 29, 31

K

Kazusa Marker Database Vi, 191, 193, 195, 196, 197, 205, 206, 287

L

Legume Crop Genome Initiative (Lcgi) 119
Legume Production 213, 223
Legumes 118
Lotus Japonicus 192, 194, 195, 196, 201, 205, 207, 209

M

Marker Assisted Selection (Mas) 42
Microarray Analysis 124

Minerals 143, 145, 146
Molecular Breeding 45, 54
Molecular Genetic 41, 42, 51, 53
Molecular Identifying Sequences (Mids) 260
Mutations 192

N

National Bureau Of Plant Genetic Resources (Nbpgr) 216
Next-Generation Sequencing (Ngs) 256, 258
Nutritious 145, 146

O

Oilseed Crop 256
Oilseed Crops 23
Oleosin 123

P

Peanut 192, 194, 197, 203, 204, 209, 210
Peanut V, 23, 26, 27, 29, 30, 31, 33, 34, 35, 36, 285
Peanut Allergy 1, 2, 9, 18, 20
Peanut Cultivar 42, 45, 52, 53
Peanut Genome 118, 119, 120, 128, 129, 132
Peanut Plants 24, 26, 27, 28, 29, 30, 31, 32
Pepper 192, 194, 197, 199
Phylogenetics 82
Plant Genomes 119, 120
Polymorphism 42, 43, 45, 51, 54
Polymorphisms 192, 207
Polyploidy 82, 83, 85, 96, 107

Q

Quality Traits 144, 154, 167, 175
Quantitative Trait Loci (Qtls) 44

R

Reduced Representation Libraries (Rrls) 256, 258
Restriction Fragment Length Polymorphism (Rflp) 192
Rich Source 145, 152

S

Saline Soil 24
Seed Storage 6
Simple Sequence Repeat (Ssr) 257
Single Nucleotide Polymorphism (Snp) , 192, 256
Soil 146, 148, 149, 155, 160, 163, 175, 186
Species 44, 45, 51, 53, 54, 56, 58
Stress Tolerance 24, 25

T

Tentative Orthologous Genes (Togs) 92, 93
Tomato 192, 194, 197, 198, 199, 204, 207, 209, 210, 211
Transgenic 45, 47, 59, 60, 62

U

Universal Network-Enabled Analysis Kit (Uneak) 259

V

Vitamins 143, 145

W

White Clover 192, 194, 203, 208
Wild Species 144, 158, 163, 165, 169, 181, 184